T0228481

International Journal
of Human–Computer Interaction

Vol. 17, No. 1   2004

Special Issue: Current Research of the Human Interface Society

Guest Editor: Osamu Katai

Contents

International Journal of Human–Computer Interaction

Published in cooperation with the International Ergonomics Association (IEA) and the Human Interface Society.

International Journal of Human–Computer Interaction is indexed or abstracted in *PsycINFO/Psychological Abstracts; ISI: Current Contents/Engineering, Computing, and Technology, Social Sciences Citation Index, Social SciSearch, Research Alert, Current Contents/Social & Behavioral Sciences; Ergonomics Abstracts; SciSearch; CompuMath Citation Index; Cambridge Scientific Abstracts: Health & Safety Science Abstracts, Risk Abstracts; INSPEC; EBSCOhost Products; Linguistics and Language Behavior Abstracts.*

First published 2004 by Lawrence Erlbaum Associates, Inc.

2 Park Square, Milton Park, Abingdon, Oxfordshire OX14 4RN
52 Vanderbilt Avenue, New York, NY 10017

Routledge is an imprint of the Taylor & Francis Group, an informa business

First issued in hardback 2019

ISSN 1044–7318
ISBN 13: 978-0-8058-9554-4 (pbk)
ISBN 13: 978-1-138-43289-5 (hbk)

INTERNATIONAL JOURNAL OF HUMAN–COMPUTER INTERACTION, *17*(1), 1–2
Copyright © 2004, Lawrence Erlbaum Associates, Inc.

Introduction

Osamu Katai

Graduate School of Informatics, Kyoto University

As you may know, our society, the Human Interface Society, has had a long association with the *International Journal of Human–Computer Interaction*, particularly through academic research activity exchanges between Professor Gavriel Salvendy and Professor Hiroshi Tamura of Hiroshima International University. This special issue is the first opportunity for us to introduce you to our society's research activities.

The rather long history of editing this issue is as follows: First, Professor Salvendy contacted the board of our society inviting us to edit a special issue on our society. Then Professor Masaaki Kurosu of Shizuoka University, now of the National Institute of Multimedia Education, who was the board member in charge of managing international relationships contacted Professor Salvendy. It has taken a rather long time to edit this issue, as the editors-in-chief of the transactions of our society have taken on the role of deciding this issue's editorial policy. This task fell to Professor Takao Kurokawa of Kyoto Institute of Technology and Professor Michiaki Yasumura of Keio University, who have been in touch with Professor Kay M. Stanney. The final stage of the edition has come to me, and the editorial policy of this issue is to edit the society's most prized articles by translating them into English, because the purpose of this issue is to introduce your readers to our research activities.

The Human Interface Society was founded in January 1999 and now has about 1,000 professional members and 150 student members. We publish journals regularly four times a year, with transactions included in each volume. Our editorial policy for the transactions is that each issue except for No. 1 and 2 of Vol. 1 should involve a "special issue subject." Each transaction involves around 10 papers and two thirds of which are on the volume's special issue subject. The following is a list of these subjects:

1999:
  Vol. 1, No. 3—Welfare Engineering
  Vol. 1, No. 4—Virtual Reality

2000:
  Vol. 2, No. 1—Human System Interactions Induced by Automation

Requests for reprints should be sent to Osamu Katai, Graduate School of Informatics, Department of Systems Science, Kyoto University, Yoshida Honmachi, Sakyo-ku, Kyoto 606–8501, Japan. E-mail: katai@i.kyoto-u.ac.jp

The list clearly shows that our research activities cover a broad range of topics, with 28 area codes as follows: Cognition; Human error; Kansei and Psychology; Physiology; Nonverbal and Multimodal; Multimedia; Network media; Agent-oriented approaches; Groupware; Communication support; Artificial reality; Approaches to the real world; Mobile and Wearable; Language, Sign, and Metaphor; Visionary disabilities; Auditory disabilities; Welfare and Support for disabilities in general; Education support; User modeling; Usability; Environmental design and Information design; Design support; Input method and Input device; Output method and Output device; Art; and Society and Transport.

The first three articles in this issue are prized articles from among the articles in the transactions from No. 1 of Vol. 1 to No. 4 of Vol. 2, whereas the second three articles are the prized papers from No. 1 of Vol. 2 to No. 4 of Vol. 3. The last article has been specially written for this issue by a special interest group, introducing research activities in Japan focusing on usability.

In closing this introduction, I extend sincere thanks to Professor Salvendy, Professor Stanney, Professor Tamura, Professor Kurosu, Professor Kurokawa, Professor Yasumura, and also a special thanks to the members of the special editorial committee for this issue: Professor Masaharu Kitamura, Management of Science and Technology Department, Tohoku University; Dr. Masaaki Kurosu, National Institute of Multimedia Education; Dr. Katsunori Shimohara, ATR Human Information Science Laboratories and Kyoto University; Professor Katsuari Kamei, Department of Computer Science, Ritsumeikan University; Dr. Michio Okada, ATR Intelligent Robotics & Communication Laboratories and Kyoto University; and Associate Professor Hiroshi Kawakami, Graduate School of Informatics, Kyoto University.

Finally, I thank the president, Professor Hidekazu Yoshikawa of Kyoto University, and the board members of our society for continually supporting this edition. We hope this issue will be interesting and valuable so that you will look forward to the succeeding special issues on our society.

INTERNATIONAL JOURNAL OF HUMAN–COMPUTER INTERACTION, 17(1), 3–24

A Microworld Approach to Identifying Issues of Human-Automation Systems Design for Supporting Operator's Situation Awareness

Makoto Itoh
Toshiyuki Inagaki
Institute of Information Sciences and Electronics
University of Tsukuba

This article gives a microworld approach to identify design requirements for better situation awareness. Two experiments were designed and conducted. In the first experiment, behaviors of operators who lost situation awareness were analyzed, and the following two findings were obtained: (a) Automation must provide an operator with feedback information on the control mode, even when it is the operator who changed a control mode. (b) Authority for control may have to be passed from an operator to automation for attaining safety in highly urgent situations. The second experiment was done to investigate how human-interface may affect an operator's situation awareness. It is shown that human-interface must be carefully designed to externalize a mental model of the controlled process.

1. INTRODUCTION

The concept of human-centered automation has been accepted widely as a promising approach to integrating humans and machines (Billings, 1997; Woods, 1989). The human-centered automation assumes that operators should bear the final responsibility for safety of a human–machine system. Correct understanding of operating conditions is essential for attaining safety of human–machine systems. It is not easy, however, to maintain correct understanding all the time, especially when a process to be controlled is large and complex, such as an aircraft and a nuclear power plant. Various accidents have been caused by operators' misunderstanding

This work has been partially supported by the Center for TARA (Tsukuba Advanced Research Alliance) at the University of Tsukuba; Grant-in-Aid for Scientific Research 07650454, 08650458, and 09650437, the Ministry of Education, Science, Sports, and Culture; the Sasakawa Scientific Research Grant from the Japan Science Society; and the First Toyota High-Tech Research Grant Program. We express our appreciation to Neville Moray for his help in the analyses of the experimental data.

Requests for reprints should be sent to Makoto Itoh, Institute of Information Sciences and Electronics, University of Tsukuba, 1-5-1 Tennodai, Tsukuba, Ibaraki 305–8573, Japan. E-mail: itoh@is.tsukuba.ac.jp

of situations; see, for example, the Three Mile Island nuclear power plant accident in 1979. How an operator's awareness of a situation may be supported appropriately in complex systems has become a vital issue (e.g., see Edwards & Lees, 1974).

Moreover, correct understanding of a situation is becoming more difficult in highly automated systems because of problems related to complexity of automation. Human operators in highly automated systems may fail to understand what the automation is doing, why it is doing it, and what it will do next (Sarter & Woods, 1995a, 1995b). In fact, accidents of highly automated systems often include lack of awareness of automated actions; see, for example, the A300-600 accident at Nagoya in 1994 (Billings, 1997).

Recently, models of *situation awareness* have been developed (e.g., see Bedny & Meister, 1999; Endsley, 1995b; Smith & Hancock, 1995). Several methods of measuring situation awareness have been proposed to evaluate the degree to which new systems or displays actually improve situation awareness (Adams, Tenney, & Pew, 1995; Endsley, 1995a). However, it is still unclear how operator's situation awareness may be supported. To enhance situation awareness, we need to clarify what factors may affect situation awareness and how.

This article investigates cognitive processes of fault management activities involving situation awareness, decision making, and implementation of actions in order to obtain requirements for human–machine system design. An experimental microworld has been developed to this aim to collect data of human operator behaviors in a complex system. A microworld is not a simulator of a real system but a representation of a physical system that reflects various physical characteristics of the real world. In other words, a microworld must be complex enough to represent various interactions that may occur in real systems, even though it is not a real system. Many useful findings have been obtained by experiments with microworlds, such as Pasteurizer (e.g., Lee & Moray, 1992), Pasteurizer II (e.g., Reising & Sanderson, 2002), DURESS (e.g., Vicente, Christoffersen, & Pereklita, 1995), and MATB (e.g., Parasuraman, Molloy, & Singh, 1993).

The structure of this article is as follows. We give a brief survey of previous studies on situation awareness and describe the objective of this article. Two experiments are conducted by using a microworld. In the first experiment, we observe and analyze operator behaviors in various scenarios and clarify requirements that are needed for supporting awareness on current modes and actions that an automatic control system takes. We also show, in the first experiment, importance of a mental model of a controlled process for achievement of situation awareness. In the second experiment, we analyze the relation between human-interface design and operators' manners of assessing situations, and show effects of human-interface design on development of mental models.

2. SITUATION AWARENESS IN SUPERVISORY CONTROL

Many modern human–machine systems are semiautonomous, where a human operator intermittently programs and receives information from a computer that interconnects the operator and the controlled process. This configuration can be represented by a human supervisory control model (Sheridan, 1992, 2002). Figure 1 illustrates a supervisory control framework.

Human operators must monitor automatic actions in order to detect failures in the controlled process. If necessary, the human operators should intervene in the automatic action and should determine and implement necessary adjustments to the automation. Correct situation awareness is vital for appropriate and timely intervention. However, it is difficult to understand correctly current status of the controlled process and computer in a complex and highly automated system. Billings (1997) showed that loss of understanding of a situation is associated with one or more automation attributes as follows: *complexity* of an automation function in the computer, *coupling* of automation functions that refers to obscure interdependence among the automation functions, *autonomy* that means real or apparent self-initiated behavior of the computer, and *inadequate feedback* given to the human operator. The complexity, coupling, and autonomy may be inevitable in an advanced automation. Thus, design of human-interface that gives operators adequate feedback from the computer is critical for supporting understanding of a situation.

Situation awareness has been attracting attention of many researchers. Endsley (1995a) defined situation awareness as perception of the elements of the environment within a volume of time and space (Level 1), the comprehension of their meaning (Level 2), and the projection of their status in the near future (Level 3). In her model, situation awareness is regarded as a state of knowledge on a dynamic environment with which humans interact. In this sense, situation awareness is a product of cognitive activity for assessing a situation or a snapshot of an operator's understanding of a situation. On the basis of the definition, situation awareness may be measured in an experiment by freezing the scenario and asking questions about the state variables (Endsley, 1995a). An advantage of this way of measuring situation awareness is that it is possible to evaluate the degree to which new technologies or design may improve or degrade operator's situation awareness. The technique also clarifies at which level of situation awareness humans need to be supported. Suppose the situation awareness measurement showed that an operator achieved situation awareness of Level 2. Then he or she would need support in projecting the future.

In Endsley's model (1995a), cognitive process for achieving each level of situation awareness is not described explicitly. Sarter and Woods (1995b) claimed that it is vital to identify factors that affect cognitive activities for situation assessment in order to support situation awareness of an operator.

Several researchers have proposed to define situation awareness as a process. Adams, Tenny, and Pew (1995) claimed that situation awareness should be seen as

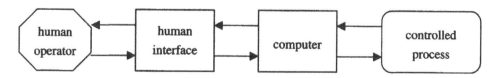

FIGURE 1 Supervisory control.

both a product and a process, on the basis of the Neisser's (1976) perceptional cycle model. Extending the idea of Adams et al. (1995), Smith and Hancock (1995) considered situation awareness to be "adaptive, externally directed consciousness" in the perception–action cycle and defined situation awareness as a generative process of knowledge creation and informed action taking. Bedny and Meister (1999) also proposed to define situation awareness as a process, based on the Russian theory of activity (Bedny & Meister, 1997). The approaches to define situation awareness as a process can explain adaptive behavior of skilled operators. That is, on the basis of operator's understanding, the operator may change his or her actions in order to acquire what he or she needs. However, it is still unclear what factors may cause failure in assessing a situation.

The objective of our study is to analyze the cognitive process so as to clarify when and how situation awareness may be improved or lost, instead of simply identifying the level of situation awareness attained. A microworld described in the next section has been developed to observe operators' behavior for fault management in a supervisory control system.

3. FLUID CIRCULATION MICROWORLD

3.1. Plant Description

The controlled plant consists of three subsystems (see Figure 2). Subsystem A is for adjusting the quality of "product fluid" for Subsystem B, which imposes the following requirements:

1. Temperature requirement: The temperature of the product fluid sent to Subsystem B must be kept within the range of 50° C to 70° C.

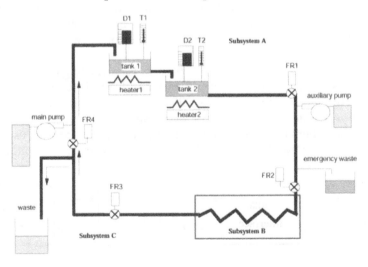

FIGURE 2 Fluid circulation microworld.

2. Flow requirement: The flow rate of the fluid to Subsystem B must be within the rage of 14 to 22, measured in an appropriate dimension.

The product fluid is utilized in Subsystem B, where the temperature of the fluid comes down to the room temperature (30° C). The used fluid goes back to Subsystem A for renewal via Subsystem C. While passing through Subsystem C, a portion of flow quantity is disposed of as waste, and the lost amount of the fluid may be compensated by the main pump at Subsystem A. Flow rate and temperature of the fluid may be adjusted at Tanks 1 and 2. The fluid level in Tank 2 must be kept within an appropriate range in order to maintain the quality of the product fluid. Energy given by each heater is assumed fixed.

Keeping the amount of fluid in Tank 2 within an appropriate range is not an easy task for a human operator. When the amount of fluid in Tank 2 increases, the temperature of the product fluid decreases, and vice versa. Adjusting the amount of fluid in Tank 2 to fulfill one of the flow and temperature requirements may cause failure to meet the other requirement. For example, suppose that too much amount of fluid has been disposed of as waste in Subsystem C. To fulfill the flow requirement, an operator needs to compensate for the lost amount of the fluid. If the operator feeds too much fluid into the tank, he or she fails to meet the temperature requirement. Feeding the appropriate amount of fluid into Tank 2 is difficult because the main pump suffers from a delay; it can start running a few seconds after the switch of the main pump is turned on.

If the flow at the outlet of Tank 2 becomes too high, an emergency waste tank located near Subsystem B may be used in order to adjust the flow rate. If the flow becomes too low, on the other hand, an auxiliary pump may be used to compensate a shortage of the product fluid. It is noted, however, that the temperature of the fluid coming through the auxiliary pump is low, at around room temperature.

Indicators are located at several places in the plant: four flow rate indicators along the pipeline (FRi, $i = 1,..,4$), and a level indicator (Di), and a thermometer (Ti) at each tank i ($i=1,2$).

3.2 Failures and Accidents

Three types of failures may occur in this plant.

Level indicator failure.   Level indicators can give an erroneous reading. The size of the reading error grows linearly as time goes on. The coefficient, ω, of the error growth with time is assumed constant during a level indicator failure. The failure itself has no effect on the quality of the product fluid, but may cause the operator's inappropriate action. For example, the operator may turn off the switch of the main pump inappropriately when the level of Tank 2 is decreasing to the lowest allowable limit. This inappropriate action may result in degradation of the product fluid quality. When the operator presses the "reset button," the level indicator becomes normal, and the reading error vanishes immediately.

Heater failure.   The temperature of the fluid at the main pump is around 30°
C. It can go up to around 50° C when heated at Tank 1, and further up to about 70° F
when heated at Tank 2. Either Heater 1 or 2 may lose capability to heat the fluid.
The heater comes back to its normal state immediately when the operator presses
the "restart button."

Pipe rupture.   Three levels are distinguished for pipe rupture: (a) the first
stage of rupture, in which 10% of flow quantity is lost, (b) the second stage of rupture, in which the loss of flow quantity becomes large exponentially as time goes
on, and (c) the third stage, in which 100% of flow quantity is lost right away. The
pipe rupture makes a transition from its first stage to the second in about 80 sec.
The second stage lasts about 30 sec, and then enters into the third stage. If the pipe
rupture of the third stage occurs at a point between Tank 2 and Subsystem B, the
flow rate to Subsystem B vanishes within 7 sec if no appropriate countermeasure is
taken. We say that an accident occurs at a point when the flow rate at Subsystem B
becomes zero.
    During the first stage of pipe rupture, the operator can "repair" it by pressing the
"repair button." If the rupture is in the second stage, the operator must activate the
auxiliary pump to compensate for the loss of flow quantity to Subsystem B; just
pressing the repair button is not enough. Once the pipe rupture enters into the third
stage, the operator must shut down the whole plant immediately to avoid an accident. It is assumed in the simulation that the repair is complete 7 sec after pressing
the repair button.

3.3 Automatic Control Systems

Three automatic controlling systems are available for a human operator: (a) an
auto-supply system that controls the main pump, (b) an *auto-compensating* system
that controls the emergency pump, and (c) an *auto-shutdown* system for shutting
down the whole plant to prevent an accident. Figure 3 shows the activation and de-
activation thresholds for those systems, on the basis of the reading of FR1 (flow indicator for the pipe between Tank 2 and Subsystem B). For example, the auto-supply system activates the main pump when the reading of FR1 goes below 17, and it
deactivates the pump when the reading of FR1 exceeds 19. The auto-compensating
system works in a similar manner. The auto-shutdown system shuts down the
plant when the reading of FR1 becomes less than 5. In this microworld, the amount
of data available for the automatic systems is limited. The automatic systems are
thus not so stupid, but not so wise.
    This microworld has been proven to be rich enough for conducting some types
of experiments on human–machine collaboration. As a matter of fact, a previous
experiment with this microworld has been successful in investigating the following issues (Inagaki & Itoh, 1998): (a) lack of situation awareness, (b) automatic control systems hiding abnormalities in the plant, (c) trust and dependence on automa-

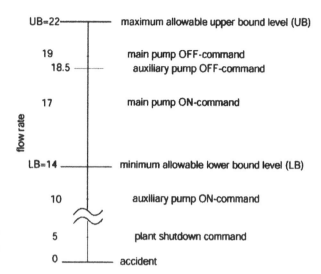

UB=22 ——————— maximum allowable upper bound level (UB)

19 — main pump OFF-command
18.5 ———— auxiliary pump OFF-command

17 — main pump ON-command

flow rate

LB=14 ———— minimum allowable lower bound level (LB)

10 — auxiliary pump ON-command

5 — plant shutdown command

0 ———— accident

FIGURE 3 Activation and de-activation thresholds for automatic control systems.

tion, (d) distrust of automation, (e) unintended use of automation, and (f) mode confusion and description error.

4. EXPERIMENT 1

4.1 Purpose

The first experiment is to see how an operator may fail to attain situation awareness when he or she works with automatic control systems. We observed operator behaviors in various scenarios and collected data showing loss of situation awareness.

We then analyzed the data on loss of situation awareness of the control modes or actions taken by automatic control systems for requirements for supporting the situation awareness.

Individual differences were investigated in the achievement of situation awareness of failures in the controlled process to show that mental models of controlled processes may affect operators' awareness of operating conditions.

4.2 Method

Tasks imposed on participants. A participant is requested to behave as an operator of the fluid circulation plant. Each operator is required to perform a "main task" and a "subtask" simultaneously.

The main task is to feed to Subsystem B the proper product fluid that satisfies the following two conditions as much as possible: (a) $50°\,C \leq$ temperature $\leq 70°\,C$, and (b) $14 \leq$ flow rate ≤ 22. To pursue the main task, the operator must (a) control the main pump appropriately, (b) decide when the auxiliary pump must be activated or stopped, and (c) decide whether the bypass line should be used for discarding the excess fluid. If some failure (pipe rupture, level indicator failure, or heater failure) occurs in the plant, the operator must take an appropriate countermeasure as soon as possible to restore the plant into its normal operating conditions. The earlier the operator performs the countermeasure, the better he or she achieves results on the main task. The most fundamental requirement imposed on each operator is "Never let an accident occur; that is, never let the flow rate at the inlet of Subsystem B become zero." If an accident occurs, the operator is regarded as having completely failed in pursuing the main task, no matter how good results he or she has achieved by the time of the accident.

In addition to the main task, every operator is requested to perform simultaneously either one of the following two subtasks: (a) to transcribe English words or sentences that are listed on several sheets of paper, which is a skill-based task, and (b) to solve problems that need reasoning, which is a rule-based or a knowledge-based task. An example of such problems for the subtask of the second kind is shown in Figure 4.

The reason for imposing subtasks on operators is that operators of real plants are usually under various types of workloads or pressure that may prevent operators from focusing their attention fully to the process control task (the main task, in this experiment). The subtask is to simulate this situation.

4.3 Participants

Seven graduate and undergraduate students participated in Experiment 1. The participants were paid some amount of money just after completion of all the trials. Each operator was informed that some monetary bonus could be given if he or she achieved either the highest or the second highest score, which is to make operators

There are seven propositions, A through G.   Three students, Rhea, Sarah,

and Tiffany have investigated the propositions.

(1) Assuming Propositions D, Rhea proved Proposition C.   Based on that,

she proved Proposition G.

(2) Assuming Proposition F, Sarah proved Proposition C.   Based on that,

she proved Proposition B.

(3) Assuming Proposition A, Tiffany proved Proposition G.   Based on that,

she proved Proposition D, and she further proved Proposition F.

The students have confirmed that every proof given in (1) through (3) is

correct.   They are now investigating propositions which are equivalent to

Proposition C.   How many propositions can they find, excluding Proposition C

itself?

FIGURE 4 An example of a problem-solving subtask.

"wise" or "ambitious," instead of letting them be "lazy," according the wording of Stassen, Johannsen, and Moray (1990). Rating of the operator's performance is done on the basis of the performance in the main task and the subtask.

4.4 Human-Interface

The participant can learn the plant status only through indicators in the control panel (see Figure 5). The control panel has been developed with the Hewlett-Packard's VEE for MS-Windows.

This display is expected to provide necessary and sufficient information for assessing the status of the controlled process and its associated automatic control systems. Operators can learn the current control mode of each automatic control system by taking a look at the mode selection switch. However, recognition of the current action of each automatic system is not straightforward. An operator can know easily that the main pump is feeding fluid if he or she finds that the reading of D1 (the level indicator of Tank 1) is becoming higher. On the other hand, status of the emergency pump or the emergency waste tank must be inferred by taking into consideration the difference between the readings of FR1 and FR2.

4.5 Experimental Design

The experiment lasted 5 days, in which it took about 1 hr each day. Table 1 shows scenarios that were used in the experiment. The scenarios are divided into four categories: (a) no failure occurs at any component, (b) a single failure occurs at a component, (c) two different components fail independently and the time between the two failures exceeds 70 sec, and (d) two or more components fail independently and the time between two consecutive failures is within 70 sec. No participant was informed, before or during the experiment, either about the categorization of scenarios or how many malfunctions might occur and when.

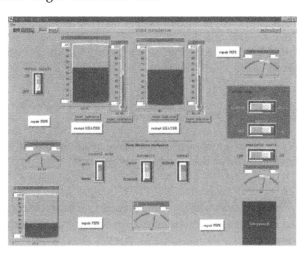

FIGURE 5 Control panel.

Table 1: Schedule for Experiment 1

			Trial			
Day	1	2	3	4	5	6
1	n	p(60)	h1(90)	p(60)	p(60)	—
2	p(150)	n	h1(100)	n	p(40)	—
3	h1(250)	p(130)	n	n	h2(180)	l(100) $\omega = +0.2$
4	h1(120), l(200), $\omega = +0.2$	h2(120), p(180)	l(70) $\omega = -0.2$, p(100)	n	—	—
5	p stage 2 (80), l(150)$\omega = +0.2$	h1(110), p(180)	l(70) $\omega = -0.1$, p(110)	n	l(130) $\omega = -0.2$, p(150), h2(180)	—

Note. n = no failure; p = pipe rupture; hi = heater i failure; l = level indicator failure. x(y)denotes the Event x occurs at time point y. ω denotes coefficient for linear growth of error in the reading of the level indicator.

Participants were requested to perform the transcribing subtask on Days 2 and 4, and problem-solving subtask on Days 3 and 5, respectively.

Participants in the experiment have full authority in determining control modes of automatic control systems. That is, a participant is allowed to engage or disengage each of the automatic control systems at any point.

4.6 Procedure

At the beginning of Day 1, each participant received oral instruction on the purpose of this experiment, the functional purpose of the plant, the structure of the plant, tasks that participants are requested to perform, possible failures in the plant, and standard countermeasure action against each failure. All trials on Day 1 and the first trial on each of the following days were given for participants to develop and maintain skill of manual control and learn what happens when a failure occurs. Participants were not given any subtask during the training phase and were not allowed to use any automatic system. Participants were given opportunities during Trial 5 on Day 1 to learn capabilities of automatic control systems for controlling the plant.

4.7 Results and Discussions

We picked up several interesting phenomena in which a participant seemed to have lost his or her situation awareness.

Mode confusion.   There were several cases in which participants misunderstood the current mode of an automatic system. Example 1 describes such an instance.

Example 1: In Trial 2 on Day 2, participant TO, who had a good mental model on the plant dynamics, engaged the auto-supply system at some point. He said at the interview after the trial, that he intended to engage the auto-shutdown system instead of the auto-supply system (see Figure 6). He also said that this was his first time performing the problem solving subtask and that he had been under pressure to cope with the new subtask. Two minutes passed before he recognized that he had engaged a wrong automatic control system. While the auto-supply system was running, participant TO sometimes turned the main pump on and off manually. While the auto-supply system is active, the main pump cannot be controlled with the main pump switch, which is a design characteristic of the plant and every participant knows that. Thus, participant TO's actions to manipulate the main pump switch were meaningless.

The reason for participant TO's error in choosing a right automatic control system may be described as follows. In the control panel, the status of the switch of the main pump does not correspond to an actual status of the main pump, and no direct information on the status of the main pump is given to operators. To find an incorrect engagement of the auto-supply system, the operator should have monitored the time-dependent behavior of readings on D1 and the position of the manual switch. However, it was not easy for participant TO to monitor all the related indicators carefully, because a given subtask was completely new to him.

From the viewpoint of human-interface design, two problems may be identified. One is that no explicit feedback is given with respect to manipulating a switch for mode selection. Such a design policy may be adopted if a designer thinks that a user takes an action on the basis of his or her intention and that there is no need to indicate explicitly which command a system has received. However, as in the case given in Example 1, a user may execute an action that does not match with his or her intention. Thus, explicit indication of commands received by the system may be effective to support operators' mode awareness of an automatic control system.

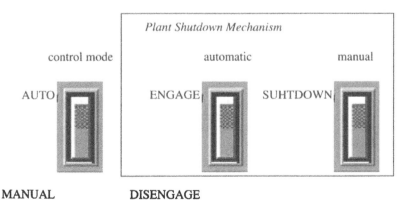

FIGURE 6 Location of switches for mode selection of the auto-supply system and the auto-shutdown system.

The second problem is that the participant receives no explicit information on what the automatic system is doing. Even though actions of the automatic system may be inferred with indirect information shown on the display, direct information may be far more helpful for an operator to improve his or her situation awareness.

Automatic feedback control hides a failure.   Table 2 shows the relation between awareness of pipe ruptures and control modes of the main pump. From this table and comments given in the interview, the following three points were identified:

1. When the plant was controlled manually, 14 out of 18 pipe ruptures (77.8%) were detected and repaired during the first stage. On the other hand, only 6 out of 31 pipe ruptures (19.4%) were detected and repaired during the first stage, when the plant was controlled by the auto-supply system. This suggests that situation awareness is likely to be lost when a human operator is out of the control-loop because the auto-supply system compensates for the loss of flow during the first stage of pipe rupture. In other words, the automatic system may hide an anomaly occurring in the pipe.

2. The auto-shutdown system prevented six possible accidents from occurring. In those cases, participants did not notice pipe rupture until the auto-shutdown system intervened to shut down the plant at the end of the second stage. This shows the effectiveness of automatic safety functions when human operators might have failed to attain appropriate situation awareness.

3. There were three accidents among 49 cases of pipe rupture. In each accident, the auto-shutdown system had been engaged until the end of the first stage of pipe rupture. When the participants noticed that the flow rate at the inlet of Subsystem B was decreasing rapidly, they disengaged the auto-shutdown system to keep the plant running. Without shutting down the plant, the participants tried to cope with the situation by activating the emergency pump manually, by sending the repair crew to fix the pipe, or both. However, the actions for keeping the plant running were too late to prevent an accident. Time available for a human operator was too short for situation assessment and decision making. This also proves the necessity of automatic safety functions under severe time criticality.

Table 2: Awareness of Pipe Rupture and Control Mode of the Main Pump

Pipe Rupture	Control Mode of the Main Pump	
	Automatic	Manual
Repaired in the first stage	6	14
Repaired in the second stage	22(6[a])	4
Accident occurred	3	0

[a]The plant was shut down by auto-shutdown system six times in 22 cases.

4.8 Individual Difference in Recognition of a Failure

Tables 3, 4, and 5 show the time elapsed before a countermeasure is taken against pipe rupture, level indicator failure, or heater failure, respectively. In most cases, participants MA and TO performed countermeasure actions against the failure at the early stage. It was found, through observation of operator behaviors and comments given in the interview after each trial, that participants MA and TO had a good mental model of the structure and mechanism of the plant as we see in Example 2.

Example 2: In Trial 3 on Day 5, a level indicator failure and a pipe rupture occurred when about 90 sec and 140 sec elapsed from the beginning of the trial, respectively. In a typical instance with an "ordinary" operator D2 failed at some point in a mode to give a reading smaller than the true value. The magnitude of reading error grew gradually with time, but the operator did not recognize that. In the meantime, a pipe rupture broke out, and FR1 gave a reading smaller than usual. D2 and FR1 happened to coincide in giving low readings. Actually, the participant thought that the coincidence suggested shortage of fluid in

Table 3: Time Elapsed Before Pipe Rupture Was Repaired

Participant	Day–Trial						
	2–5	3–2	4–2	4–3	5–2	5–3	5–5
MA	34	35	5	2	6	18	2
TO	11	68	14	22	23	57	28
IN	36	64	29	75[b]	5	65	64
SH	34	72	46	72	64	74	66
TS	78[a]	89[b]	65	75[a]	66	66	63
KA	25	90[b]	72[a]	71	74	27	63
MI	76	63	73[a]	68	69[a]	89	67[b]

[a]Auto-shutdown system shut down the plant at that time point to prevent an accident. The participant did not perform actions to repair the pipe. [b]An accident occurred.

Table 4: Time Elapsed Before Failed Level Indicator Was Reset

Particpant	Day–Trial			
	3–6	4–3	5–3	5–5
MA	42	18	30	24
TO	49	48	88	43
IN	108	75[a]	226	81
SH	172	181	226	80
TS	135	105[a]	192	56
KA	137	168	220	98
MI	53	173	129[b]	98

[a]Auto-shutdown system shut down the plant at that point to prevent an accident due to the pipe rupture after the level indicator failure. The participant did not perform actions to repair the pipe. [b]Accident occurred.

Table 5:   Time Elapsed Before Failed Heater Was Restarted

Participant	Day–Trial				
	2–3	3–5	4–2	5–2	5–5
MA	24	19	32	31	22
TO	53	15	16	28	12
IN	56	43	43	34	61
SH	75	41	52	71	74
TS	44	41	56	96	30
KA	33	39	122	58	45
MI	15	21	66	82	30

Tank 2. He should have taken seriously the phenomenon that the level of Tank 1 was going up at that time: The auto-supply system had detected that the flow rate was too low and turned on the main pump. The participant finally noticed the pipe rupture after it evolved to the second stage in which the loss of flow quantity became large. He did not recognize the level indicator failure for about three more minutes after detecting the pipe rupture.

"Smart" participants, MA and TO, on the other hand, found the pipe rupture and the level indicator failure almost immediately in the same circumstance. When a smart participant found the simultaneous decrease in readings of FR1 and D2, he compared the reading of FR1 with those of indicators FR2 and FR3 at downstream. Because of time delay, the effect of the pipe rupture had not reached FR2 and FR3 yet at that time. The participant thought that pipe rupture was occurring at somewhere upstream of FR1 and pressed the pipe repair button. Moreover, he found the failure of D2 based on his knowledge of the relationship between the fluid level in the tank and the flow rate. He thought that the reading of Level Indicator 2 was "too low," judging from the reading of FR2.

5. EXPERIMENT 2

5.1 Purpose

The results of Experiment 1 suggest that the operator's mental model of the plant dynamics strongly affects his or her awareness of the ongoing situation. Experiment 2 investigates how human-interface may affect mental models of a controlled process, in which two design alternatives were given for layout of indicators.

5.2 Method

Tasks imposed on participants.   Participants were requested to perform only the main task in the fluid circulation plant in order to analyze effects of human-interface design on mental models of the plant. The level of performance of the main task for each trial was scored according the cumulative amount of the product flow that satisfies the temperature and flow requirements. Maximum possible score for each trial was 6,000, and minimum possible score was 0. Each participant was informed how the performance was evaluated.

Participants.   Sixteen graduate and undergraduate students participated in Experiment 2. Participants were paid some amount of money, depending on their performance just after completion of the experiment. Participants were told that they could win bonuses for each trial, in addition to the base payment, if the performance score was greater than 90% of the maximum score.

Experimental design.   Two types of control panels, Interface A (see Figure 7), and Interface B (see Figure 8), were compared in this experiment. Contents are the

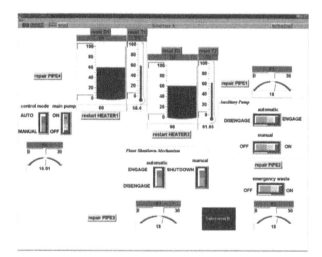

FIGURE 7 Interface A.

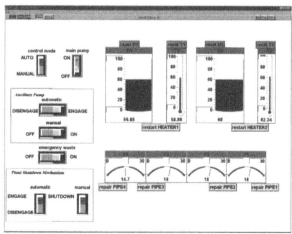

FIGURE 8 Interface B.

same in both panels. The difference between the two panels lies only in the layout of gauges and control switches. In Interface A, positions of indicators and control switches are determined based on physical locations of those items in the plant. Interface A was designed to support interpretation of variables of components by visualizing physical relations among components. In other words, Interface A may be useful to attain situation awareness of Level II. In Interface B, on the other hand, items of the same types are gathered. Interface B looks simple and makes it easier to compare readings of indicators. Interface B may be useful to attain situation awareness of Level 1. A hypothesis to be investigated is that Interface A supports the operator's situation awareness better than Interface B. Participants are randomly divided into two groups: (a) Group A to be exposed to Interface A and (b) Group B to be exposed to Interface B.

Procedure.   The experiment lasted 3 days, about 1 hr each day. At the beginning, a participant was told the purpose and the procedure of the experiment. Each participant received some training trials (at least five times) to develop manual skill for controlling the plant. No failure was assumed to occur in the training trials. During the training trials, three questions were given to each participant at the end of each trial to check whether he or she understood properly the mechanism and dynamics of the plant. The questions were on (a) time elapsed for fluid to travel from the outlet of Tank 2 into the inlet of Subsystem B, (b) time elapsed for starting the main pump after the main pump switch was turned on, and (c) mathematical relation between the level of a tank and the flow rate at the outlet of the tank. A participant was allowed to proceed to the data collection phase only if he or she achieved more than 80% of the maximum performance in every one of the last three trials and if he or she could correctly answer the aforementioned questions.

Each participant received 10 trials for data collection. Among those trials, 6 of them included a failure (F1 to F6), and the remaining 4 trials had no failure. The six failures are described as follows.

- F1: Level indicator of Tank 1 (D1) started to give an erroneous reading when 286 sec elapsed from the beginning of a trial. The coefficient, ω, of the error growth with time in the reading of D1 is +0.2.
- F2: Level indicator of Tank 2 (D2) gave an erroneous reading when 169 sec elapsed. The coefficient ω is –0.2.
- F3: The fluid began to leak at a site between Tank 2 and FR1 when 39 sec elapsed.
- F4: Heater 1 failed when 286 sec elapsed.
- F5: Heater 1 failed when 234 sec elapsed.
- F6: Heater 2 failed when 169 sec elapsed.

Table 6 depicts the experimental schedule.

Measures.   Manual control skill of an operator was evaluated on the basis of the performance score. The operator's situation awareness was measured in terms

Table 6:   Schedule in Experiment 2

Participant	Trial									
	1	2	3	4	5	6	7	8	9	10
Group A										
s1	F5	F3	N	N	N	N	F2	F4	F1	F6
s2	F4	N	F2	N	F3	F6	F1	N	N	F5
s3	F3	F5	F6	F4	N	F2	N	N	F1	N
s4	F4	F2	N	N	N	F3	F1	F5	N	F6
s5	F6	F5	F1	N	N	F4	N	N	F2	F3
s6	F2	N	F4	N	F1	F6	N	F3	N	F5
s7	F2	F4	N	N	F3	F6	N	F5	F1	N
s8	F4	N	F1	N	F6	N	F3	N	F2	F5
Group B										
s9	F6	F3	F1	F5	N	F2	N	N	N	F4
s10	N	F3	F1	F2	F5	F4	N	F6	N	N
s11	N	F4	F3	F2	N	F6	N	F1	F5	N
s12	F3	N	N	F6	F4	F5	N	N	F2	F1
s13	N	N	N	F2	F5	N	F4	F3	F6	F1
s14	F3	F5	N	F6	N	N	N	F4	F1	F2
s15	F3	N	F2	N	F1	F6	F4	F5	N	N
s16	F3	F2	F4	N	N	F1	N	F6	N	F5

of the time to detect a failure, although the time to detect a failure is an indirect index of situation awareness (Endsley, 1995a).

5.3 Results and Discussion

Acquiring skills for controlling the plant.   Figure 9 depicts learning curves for the acquisition of skills for controlling the plant. The left half portion depicts the result of Group A and the right half shows for Group B. Performance scores of both

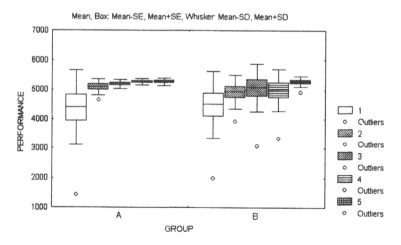

FIGURE 9 Learning curves for acquisition of controlling skills.

groups seem to converge around 5,300. It can be said that almost all the participants acquired skills that are good enough to control the plant in normal operating conditions. However, the speed of acquiring manual control skill with Interface A seemed to be faster than that with Interface B.

Fault management.   Figure 10 shows the cumulative rates of participants who succeeded in detecting a failure. Among six failures, four (F1, F4, F5, and F6) out of them reveal that the time to detect a failure was shorter under Interface A than under Interface B.

Now let us test the difference of the time to detect a failure between Interfaces A and B. Because there are several censored data that show some participants failed to find a failure by the end of the trial, the analysis of variance (ANOVA) may not be done in a straightforward manner. The true values for the censored data can be inferred in several ways. For example, one approach is assuming that participants in the censored cases would find the failure immediately after the end of the trial (the "best" estimation), and another approach is assuming that the participants shall not recognize the failure forever (the "worst" estimation). We assess the censored data, in this article, on the basis of the time-dependent human reliability model (Dougherty & Fragola, 1988). In the model, the censored data is assumed to be located somewhere on the log-normal plot (e.g., see Figure 11), in which two parameter values, t.5 (the median of the elapsed time to detect the failure) and t.95 (the 95th percentile response time to detect the failure), determine the shape of the log-normal plot. Table 7 gives the values of t.5 and t.95 for the log-normal plots. Data on F3 has been removed because the nature of the failure in F3 was different from other failures: A trial under F3 terminated immediately when an accident occurred due to a pipe rupture. No significant group effect was observed through one-way ANOVA on those parameters.

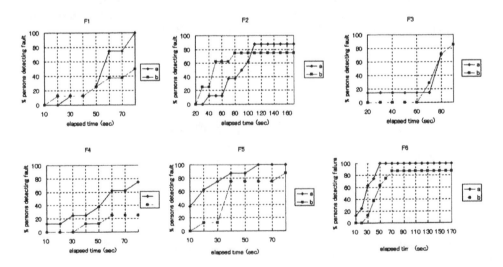

FIGURE 10 Cumulative plots of time to detect a failure.

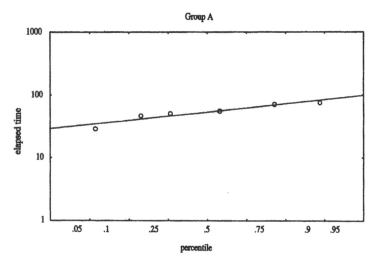

FIGURE 11 An example of log-normal plots of time to detect a failure (Group A, F1).

The previous results of ANOVAs suggest that difficulty in recognizing a failure with a control panel depends on the types of failures. From Table 7, it is said that Interface A supports better recognition of F1, F4, F5, and F6 than Interface B. On the other hand, Interface B is better than Interface A for recognizing F2 for some participants. To clarify the reason for the dependence of difficulty in recognizing a failure on the types of failures, we analyzed how failure detection proceeded for each type of failure.

1. Recognition of a level indicator failure (F1 and F2). Three kinds of differences may be used as cues for finding F1 failure: (a) difference between the readings of D1 and D2, (b) difference between the readings of D1 and T1, and (c) difference between the readings of FR4 and D1. Similarly, the following three types of differences in sensor readings may be used as cues for finding F2 failure: (a) difference between the readings of D1 and D2, (b) difference between the readings of D2 and T2, and (c) difference between the readings of D2 and FR1. If an operator paid the equal amount of attention to both Tanks 1 and 2, the cumulative plots of time to de-

Table 7: Parameters of Log-Normal Plots

Failure	A		B	
	t.5	t.95	t.5	t.95
F1	54	88	111	1173
F2	80	223	49	153
F4	61	697	157	1300
F5	16	64	42	90
F6	26	63	45	98

tect a failure in D1 and D2 would be the same. However, Figure 10 shows that the amount of attention that a participant paid differs from tank to tank. For example, for Group A, recognizing a level indicator failure in D1 is easier than in D2. Figure 10 also shows that Groups A and B differ in a way of paying attention to tanks. For F1, all participants in Group A recognized a failure in D1 but some participants in Group B failed to find the failure by the end of the trial. For F2, on the other hand, many participants in Group B found a failure in D2 earlier than participants in Group A. That is, monitoring policy may be affected by the human-interface (the layout of indicators).

2. Recognition of a pipe rupture (F3). Because the quantity of loss is small during the first stage of pipe rupture, it may be difficult to notice the rupture under both Interfaces A and B. Once the rupture enters into the second stage, on the other hand, the readings of the flow rate indicators at downstream decrease drastically. Using the rapid decrease of reading as a cue, the pipe rupture can be found easily under both Interfaces A and B. Thus, difficulty in recognizing a pipe rupture is independent of the layout of the interface.

3. Recognition of a heater failure (F4, F5, and F6). The following two cues may be useful for finding the heater failure: (a) difference between readings of mass and temperature in the tank, (b) extremely low temperature at the tank. The first cue can be used at the early stage of the heater failure. The second cue, however, may not be noticeable until the temperature goes down to a sufficiently low level, say, the lowest allowable limit ($50°$ C). Figure 10 shows that the participants in Group A are better than those in the Gro up B at recognizing a heater failure for each of the cases, F4, F5, and F6. This result implies that many participants in Group A used the first cue, and that they had a good mental model of the plant.

The result mentioned in Number 3 suggest that operators' mental models of the plant dynamics depend on the layout of the human interface.

6. CONCLUSIONS

Processes of assessing operating conditions were analyzed in this article to clarify requirements for human–machine system design to support situation awareness in fault management tasks. We obtained the following three points that are vital in the design of human-interface and responsibility allocation between humans and automation:

1. An automated control system must inform operators clearly of its control mode and what it is doing. It is important to provide the operator with feedback information on the control mode, even when it is the operator who changed a control mode. This is because a mode change command may be sent to the automatic control system unintentionally.

2. In some cases, actions taken by an operator were not effective for attaining safety. The authority for controlling the plant may have to be passed from an operator to automation in order to avoid accidents in highly urgent situations. This find-

ing supports the idea of situation-adaptive autonomy (Inagaki, 2003), where the authority and responsibility of tasks are traded between humans and automation dynamically depending on the situation.

3. Operators' mental models of a controlled process was heavily affected by human-interface design. In other words, the human interface must be carefully designed to externalize a mental model of the controlled process for better support of situation awareness of Level 2 or 3.

The previous points are also effective for acquiring human trust in an automatic control system. Acquiring human trust is vital in human-centered automation (Sheridan, 1997), and is especially important in situation-adaptive automation (Inagaki, Moray, & Itoh, 1998; Moray, Inagaki, & Itoh, 2000).

Further study is needed to prevent overreliance on automatic control systems. Every automatic control system has limitations. Information on the limitation must be shown explicitly to the operators in order to reduce accidents due to overreliance that may occur when a human operator trusts automation even under a condition that is beyond the capability of the automation (Itoh, Inahashi, & Tanaka, 2003).

REFERENCES

Adams, M. J., Tenny, Y. J., & Pew, R. W. (1995). Situation awareness and the cognitive management of complex systems. *Human Factors, 37*, 85–104.

Bedny, G., & Meister, D. (1997). *The Russian theory of activity*. Mahwah, NJ: Lawrence Erlbaum Associates, Inc.

Bedny, G., & Meister, D. (1999). Theory of activity and situation awareness. *International Journal of Cognitive Ergonomics, 3*, 63–72.

Billings, C. (1997). *Aviation automation*. Mahwah, NJ: Lawrence Erlbaum Associates, Inc.

Dougherty, E. M., & Fragola, J. R. (1988). *Human reliability analysis*. New York: Wiley.

Edwards, E., & Lees, P. (Eds.) (1974). *The human operator in process control*. London: Taylor & Francis.

Endsley, M. (1995a). Measurement of situation awareness in dynamic systems. *Human Factors, 37*, 64–84.

Endsley, M. (1995b). Toward a theory of situation awareness in dynamic systems. *Human Factors, 37*, 32–64.

Inagaki, T. (2003). Automation and the cost of authority. *International Journal of Industrial Ergonomics, 31*, 169–174.

Inagaki, T., & Itoh, M. (1998). Trust, autonomy, and authority in human–machine systems: Situation-adaptive coordination for systems safety. In E. Hollnagel & H. Yoshikawa (Eds.), *Cognitive Systems Engineering for Process Control '96* (pp. 19–28). Groningen, The Netherlands: European Society for the Study of Cognitive Systems.

Inagaki, T., Moray, N., & Itoh, M. (1998, September). Trust, self-confidence and authority in human–machine systems. *Proceedings of IFAC Man-Machine Systems, 431–436.*

Itoh, M., Inahashi, H., & Tanaka, K. (2003, August). Informing limit of ability and its reason for reducing overtrust in automation. *Proceedings of XVth Triennial Congress of the International Ergonomics Association, 1*, 141–144.

Lee, J., & Moray, N. (1992). Trust, control strategies and allocation of function in human–machine systems. *Ergonomics, 35*, 1243–1270.

Moray, N., Inagaki, T., & Itoh, M. (2000). Adaptive automation, trust, and self-confidence in fault management of time-critical tasks. *Journal of Experimental Psychology: Applied, 6*, 44–58.

Neisser, U. (1976). *Cognition and reality: Principles and implications of cognitive psychology.* San Francisco: Freeman.

Parasuraman, R., Molloy, R., & Singh, I. L. (1993). Performance consequences of automation-induced "complacency." *The International Journal of Aviation Psychology, 3*, 1–23.

Reising, D. C., & Sanderson, P. (2002). Work domain analysis and sensors 2: Pasteurizer II case study. *International Journal of Human-Computer Studies, 56*, 597–637.

Sarter, N. B., & Woods, D. D. (1995a). Automation surprises. In G. Salvendy (Ed.), *Handbook of human factors and ergonomics* (pp. 1926–1943). New York: Wiley.

Sarter, N. B., & Woods, D. D. (1995b). How in the world did we ever get into that mode? Mode error and awareness in supervisory control. *Human Factors, 37*, 5–19.

Sheridan, T. B. (1992). *Telerobotics, automation, and supervisory control.* Cambridge, MA: MIT Press.

Sheridan, T. B. (1997). Supervisory control. In G. Salvendy (Ed.), *Handbook of human factors and ergonomics* (pp. 1295–1327), New York: Wiley.

Sheridan, T. B. (2002). *Humans and automation: System design and research issues.* Santa Monica, CA: Wiley.

Smith, K. & Hancock, P. (1995). Situation awareness is adaptive, externally directed consciousness. *Human Factors, 37*, 137–148.

Stassen, H., Johannsen, G., & Moray, N. (1990). Internal representation, internal model, human performance model and mental workload. *Automatica, 26*, 811–820.

Vicente, K., Christoffersen, K., & Pereklita, A. (1995). Supporting operator problem solving through ecological interface design. *IEEE Transactions on Systems, Man, and Cybernetics, 25*, 529–545.

Woods, D. D. (1989). The effect of automation on the human's role: Experience from non-aviation industries. In S. Norman & H. Orlady (Eds.), *Flight deck automation: Promises and realities* (NASA-CP10036, pp. 61–85). Moffett Field, CA: NASA-Ames Research Center.

INTERNATIONAL JOURNAL OF HUMAN–COMPUTER INTERACTION, *17*(1), 25–42

Evaluation of Organizational Structure in Emergency Situations From the Viewpoint of Communication

Shogo Nishida
Takashi Koiso
Mie Nakatani
Department of Systems Innovation, Graduate School of Engineering Science
Osaka University

This article focuses on evaluation of organizational structure in emergency situations from the communication viewpoints. The communication process in an emergency is analyzed first, then the problems caused in the process are discussed. A communication model is proposed that considers human related factors such as an organization's "competence," "duty," "responsibility," and "knowledge." On the basis of the model, a system to evaluate organizational structure in emergency situations from the viewpoint of communication is designed. Finally, a prototype system is developed, and its evaluation results are represented. The article closes with a discussion of how the proposed communication model could be used to aid in the development of group communication support systems for emergency situations.

1. INTRODUCTION

It is frequently observed in emergency situations that important information does not reach the appropriate person or department in an organization because of confusion after the emergency or in some cases because of lack of a contact address. This phenomenon happens especially when the size of an organization becomes large, and we believe it is very important to predict communication problems in advance and improve the organizational structure to manage emergency situations from the viewpoint of communication.

This article deals with the evaluation of organizational structure in emergency situations from the communication viewpoints. Considerable research has been

This research was partly supported by the Japan Society for the Promotion of Science under Grand-in-Aid for Creative Scientific Research (Project No. 13S0018).

Request for reprints should be sent to Shogo Nishida, Dept. of Systems Innovation, Graduate School of Engineering Science, Osaka University, 1–3 machikaneyama, Toyonaka Osaka 560–8531, Japan. E-mail: nishida@sys.es.osaka-u.ac.jp

conducted on communication support system in the field of Computer-Supported Cooperative Work (CSCW). Among them are teleconference systems (Ishii, 1990; Kawabe & Sakata, 1990; Okada, Ichikawa, Jeong, Tanaka, & Matsushita, 1995), information filtering systems (Malone, Grant, Lai, Rao, & Rosenblitt, 1986), and group decision support systems (Malone, 1987; Nunamaker, 1989). Our system is based on a communication model that considers human related factors in an organization during emergency situations. Several such communication models have been proposed in the field of CSCW. For example, a conversation model based on the Speech Act Theory was developed (Winograd, 1988) and has been used as a support system that deals with e-mail processing. The Issue Based Information System model was proposed (Conklin & Begeman, 1988) and has been used as a support system to enhance software productivity. Furthermore, a trouble communication model for supporting a software development project was studied by Nakatani and Nishida (1992).

In this article, first communication in emergency situations in a hierarchical management structure for large scale systems is analyzed and the important factors in emergency situations are analyzed. Next, a communication model in emergency situations is proposed, which focuses on human related factors such as an organization's "competence," "duty," "responsibility," and "knowledge." Then a system to evaluate organizational structure in emergency situations from the viewpoint of communication is proposed based on the model. Finally, a prototype system is developed, and its evaluation results are presented. Furthermore, development of a group communication support system for emergency situations based on the communication model is also discussed.

2. ANALYSIS OF COMMUNICATION IN HIERARCHICAL ORGANIZATIONS IN EMERGENCY SITUATIONS

Recently, "commandware" has been recognized as very important in the field of crisis management (Kawata, 1995). Commandware is regarded as the chain of commands required to manage emergency situations. Commandware is closely related to the structure of an organization. Hierarchical structures are usually adopted for management of large scale systems such as fire departments, police systems, management systems for large scale chemical plants, and so on. In such hierarchical structures, problem solving is conducted by mutual communication among nodes in the hierarchy as shown in Figure 1.

We held interviews with operators of a power plant and firemen from a fire department to investigate how decisions or judgments are conducted and what types of communications occur in emergency situations. We asked questions regarding the process of management decision making in power plants and fire departments. Figure 2 shows a judgment process model in each node of the hierarchy in an emergency, which was derived from the interviews.

This judgment model consists of part "recognition of situation" and part "decision of action." In the former part, a current situation is recognized based on the information gathered from other nodes and one's own information. In the latter part,

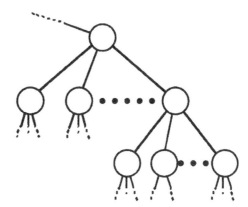

FIGURE 1 Hierarchical organization.

proper actions are selected from action candidates, which are derived from the current situation by taking some kinds of restrictions into consideration. Information gathering is conducted also in this phase. In the previous judgment process model, considerable communication occurs among the other nodes in the hierarchy.

Next, interviews were conducted that focused on viewpoint of communication. Concretely, we asked what type of communications occurred and what the objectives of the communication were. By summarizing these data, the conclusion was reached that much of the communications observed in emergency management are caused by the following factors:

1. Who is the person that has competence to execute the operation?
2. Who is the person that must execute the operation?
3. Who is the person that takes responsibility for assuring some results in the troubled situation?
4. Who is the person that has knowledge of the current situation or of the actions to be taken?

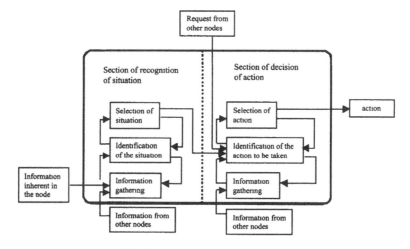

FIGURE 2 Judgement process model.

We call the previous factors "human related factors," and it is thought that they play a very important role in selecting proper actions in the judgment process model. Human related factors are the causes of communications between each node in a hierarchical organization, and the contact address as well as the quantity of communication are determined by human related factors. Although little attention has generally been given to the communication caused by these factors, we recognized the importance of human related factors through the analysis of the interviews with plant operators and fire department personnel. In the next section, an emergency communication model is discussed based on human related factors.

3. A COMMUNICATION MODEL IN EMERGENCY SITUATIONS

3.1. Types of Communication in Emergency Situations

Here, four types of communication generated at each node of a hierarchical organization are considered, as shown in Figure 3.

1. Communication generated by competence: This is defined as the communication to get permission for executing some operation, because the requesting person has no competence, that is, no right to execute the operation. For example, Operator A has a competence to stop a plant by activating the main switch. In this case, if Operator B, who has no such competence, wants to stop the plant, then Operator B has to communicate with Operator A in order to get permission to stop the plant.

2. Communication generated by duty: Duty is defined as a job that has to be executed by some person under a given situation. Therefore, communication generated by duty is related to another person who has the duty to execute some operation under a given situation, because the current situation is thought to need the operation and the requesting person has no duty to execute it. For example, Operator A has a duty to shut down a valve if a temperature sensor indicates more than 200°. In this case, if Operator B finds that the temperature sensor indicates 230°,

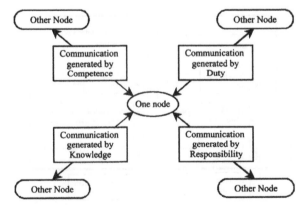

FIGURE 3 Communications in emergency situations caused by human related factors.

then Operator B needs to communicate the situation to Operator A to urge him to shut the valve down.

3. Communication generated by responsibility: Responsibility is defined as a guarantee of some result by taking any means under a given situation. Communication generated by responsibility is a communication to contact another person who has the responsibility to execute some operation for assuring some results, because the requesting person has no responsibility for it. For example, Operator A has a responsibility to avoid causing a fire associated with the temperature rise of a tank. In this case, if Operator B finds a temperature rise of the tank, he needs to communicate with Operator A to communicate the situation, because Operator A has to take any means to prevent fires.

4. Communication generated by knowledge: This is defined as the communication to contact another person who knows the situation or the operation very well, because the requesting person does not know it well. For example, Operator C has deep knowledge on the control panel of a plant. In this case, if Operator A is a novice and experiences trouble with operation of the control panel, then he should communicate with Operator C to get some advice or knowledge on the troubled situation.

Our communication model of emergency situations consists of the previous four types of basic communications. The model is composed of both the "physical structure" of a large scale system and its associated "human related factors."

3.2. Components of the Communication Model

Figure 4 shows the concrete components of physical structure and human related factors. Physical structure, which is decided by the structure of artifacts, is divided into four sub structures as follows:

1. Plant structure (PS): PS shows the name and layer structure of each part of the target system.

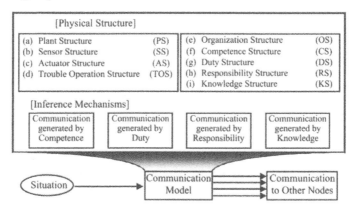

FIGURE 4 Communication model.

2. Sensor structure (SS): SS shows the sensor name, location, measurement, and values.
3. Actuator structure (AS): AS shows the actuator name, location, and type of actuator.
4. Trouble operation structure (TOS): TOS shows the relation between the emergency situation, action to be taken, and result of the action.

These substructures are determined by the physical factors of the system and are independent of the human related factors. On the other hand, human related factors are concerned with competence, duty, responsibility, and knowledge. These latter factors are determined by the formation of an organization, bylaws related to management, knowledge of each member of the organization, and so on. The human related factors are expressed by the following five substructures:

1. Organization structure (OS): OS shows the name of each member of the organization and its layer structure.
2. Competence structure (CS): CS indicates the relation between members of the organization and their competence. Concretely, the name of each member and actions to be taken are assigned as the CS data.
3. Duty structure (DS): DS indicates the relation between members of the organization and their duty. Concretely, the name of the member, the actions to be taken, and the situation are assigned as the DS data.
4. Responsibility structure (RS): RS indicates the relation between members of the organization and their responsibility. Concretely, the name of the member, situation, and result to guarantee are assigned as the RS data.
5. Knowledge structure (KS): KS indicates the relation between members of the organization and their knowledge. Concretely, the name of the member and target of knowledge are assigned as the KS data.

3.3. Inference Mechanism to Decide Communication

Adding these structural data, the model has an inference mechanism to predict communication generated by the four types of human related factors. When some situation is given as an input, the contact person of each type of communication is determined by the following mechanisms based on the physical structure and the human related factors.

1. Communications generated by competence: Some appropriate actions to be taken for the input situation are selected from TOS data. Then persons who have competence for the selected actions are predicted from the CS data, and the contact person of communication generated by competence is decided.
2. Communications generated by duty: Some appropriate actions to be taken for the given situation are selected from TOS data. Then persons who have duty regarding the selected actions for the current situation are calculated from DS data.

The derived persons correspond to the contact address of communication generated by duty.

3. Communications generated by responsibility: Some appropriate actions to be taken and results of the actions under the given situation are selected from TOS data. Then persons who have responsibility to assure some results for the current situation are predicted from RS data. The derived persons correspond to the contact address of communication generated by responsibility.

4. Communications generated by knowledge: Some parts of the plant structure concerned with the given situation or some appropriate actions to be taken under the given situation are selected from PS, SS, AS, and TOS data. Persons who have knowledge of the part selected by PS, SS, and AS are calculated from KS data. Moreover, persons who have knowledge of the selected actions are also calculated from KS data. The contact persons of communication generated by knowledge are decided in this way.

As a total, the communication model in emergency situations can predict a contact address by symbol processing using data of both physical structure and the human related factors when the situation of the system is given as an input.

3.4. An Example of Simulation

A communication simulator was developed based on the previously mentioned communication model. The function of this communication simulator was to predict types of communication and their route when input was given to the simulator on some situation.

The data of the simulator are managed by frame-type data as shown in Table 1. Table 1 shows the concrete data structures for PS and CS. Other types of data structures used in the communication model are the same as the one shown in the table. For example, PS data (pnm, 1, Pa, In-s) means that the part "n-s" is in the plant; "Pa" is defined as "No.1" plant structure data. On the other hand, (cmp, (adj, 1, U), A) means that Person A has a competence to operate Actuator 1 to U(p). By using these nine types of data, the destination of each kind of communication, which is generated from competence, duty, responsibility and knowledge, is determined by the mechanism mentioned in Section 3.3.

Table 1: Data Structure in Prototype System

Plant Structure (PS)	Competence Structure (CS)
(pnm,(=flag for PS),	
(component number),	(cmp,(=flag for CS),
(name of the first layer part),	(action to be taken),
(name of the second layer part),	(name of the person))
(name of the n-th layer part))	
Example: (pnm, 1,Pa,In-s)	(Example: (cmp, (adj,1,U,A)

Figure 5 shows example data for the physical structure and human related factors of a chemical plant management case. The structure of organization is assumed to be a three-layer hierarchy, which has seven operators as a total. The physical structure is composed of four plants and their connection parts, and each plant has several sensors and actuators. Parameters of sensors take values of "H(igh)," "M(iddle)," and "L(ow)." Parameters of actuators take values of "U(p)" and "D(own)" in this case.

The input for the system is the situation of the plant, and it is expressed by the indication values of each sensor. An example of the simulation result is shown in Figure 6, where input situation is assumed to be (sen, 10, H), which means that Sensor 10 becomes High, and the finder of the situation is F. As shown in the figure, four types of communication are derived from the input data, and the communication route is also predicted.

The detailed process of the calculation for derivation of the communication is as follows. At first, the input (sen, 10, H) is checked with the situation part of each TOS data and both "action to be taken" and "result after the action" are calculated from

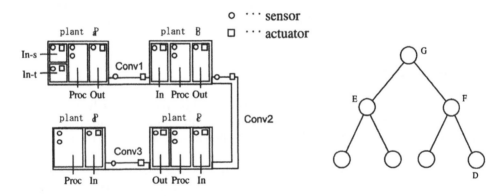

Physical structure

FIGURE 5 Example data: A chemical plant.

Input: Given situation : (sen,10,H)   Finder of emergency : F
Output: Communication generated by Competence :
            F → E (Case 1)   F → B (Case 2)
            Communication generated by Duty :
            (No Communication)   (Case 1) F → E (Case 2)
            Communication generated by Responsibility :
            F → B (Case 1)   F → B (Case 2)
            Communication generated by Knowledge :
            F → E (Case 1)   F → G (Case 2)

            Case 1 : when Actuator 5 is operated
            Case 2 : when Actuator 6 is operated

FIGURE 6 An example of simulation results.

the matched TOS data. In this case, selected "action to be taken" and "result after the action" are:

Action to be taken: adj, 1, U
Result after the action: sen, 1, M

Next, the name of the person related to each communication is calculated from human related factors data and the above selected data. In the case of competence, the "action to be taken" part of TOS is checked with that of CS, and Person E and B are selected depending on the operated actuator. In the same way, contact persons of other types of communications are obtained as shown in the figure.

4. DESIGN OF EVALUATION SYSTEM FOR ORGANIZATIONAL STRUCTURE IN EMERGENCY

One of the possible applications of the proposed communication model is to build an evaluation system of organizational structure for emergency situations from the viewpoint of communication. In reality, the structures of competence, duty, and responsibility are defined by the organization in advance. Because the proposed communication model is separated into two types of data, physical structure and human related factors, communication patterns can be calculated for various types of input situations by changing the human related factors of an organization, such as competence structure, duty structure and so on, whereas the physical structure remains fixed.

Prediction of the communication pattern for the current organization will be useful for finding communication problems when an emergency occurs. Furthermore, predictions of communication patterns when organizational structure is changed will be useful for organizational design. Design issues of evaluation systems for these purposes, especially concrete evaluation indexes, are discussed in the next section.

4.1 Index on Communication Bottleneck

It is frequently observed that communication is concentrated on some person or some section of an organization in emergency situations, and it causes problems such as slow-down of communication, delay of decision making, and so on. Here a communication bottleneck is defined as the person that has a high possibility of receiving many communications during an emergency. The communication bottleneck is predicted by calculating the number of communications that reach each person during various types of situations using the communication simulator. The number of communications to each person is calculated by using the occurrence possibility of all types of situations and communication routes predicted by the simulator for each situation. From these data, a distribution pattern of communication frequency is obtained, and the person who has a relatively high value is assumed to be a communication bottleneck.

4.2. Index on the Balance Between Assigned Human Related Factors and Used Ones

Human related factors such as competence, duty, and responsibility are determined for each member in the organization in advance. It is possible, however, that some of the human related factors are not used in real situations, because the intension of the founder of the organization is not reflected exactly or because the structure of the organization has changed. To detect this situation, an "index of the balance between assigned human related factors and used ones" is proposed for the evaluation system. The index is calculated based on the number of each human related factors assigned in advance and the number of communications generated from each human related factor. For example, the number of competence, duty, and responsibility assigned in the organization is 100,100,50 and the number of communications generated from competence, duty, and responsibility is 180,10,40. In this case, the communication frequency derived from duty structure is relatively low compared with other human related factors. Such an indication of imbalance between both data is thought to be useful for the manager of an organization. This index can be calculated both for each member and for the whole organization.

4.3 Index on Communication Extension

It is very important in emergency situations that communication is limited to some small number of members of the organization in order to speed up decision making and to avoid troubles in communication. Here the following two indexes are used to detect extension of the communication.

Index of Horizontal Extension (Eh).   Index of Horizontal Extension (Eh) is defined by the following expression, which indicates the average distance of related nodes in the horizontal direction.

$$Eh = \frac{\displaystyle\sum_{j=1}^{N} \sum_{i=1}^{N} Dist(i, j)}{{}_N C_2} \tag{1}$$

Where N: total number of related nodes for some communication
　　NODE$_i$: ith node number (i = 1~ N)
　　Dist(i,j) : distance between NODE$_i$ and NODE$_j$
　　$_n C_2$: n(n-1)/2

Eh value for Nodes C,E,F in Figure 5b becomes 2.00 from Equation.

Index of Vertical Extension (Ev).   Index of Vertical Extension (Ev) is defined by the following expression, which indicates the average distance of related nodes in the vertical direction.

$$Ev = \frac{\sum_{j=1}^{N} \sum_{i=1}^{N} E(i,j)}{_N C_2} \qquad (2)$$

Where E(i,j) : =hd(i,j) when Dist(NODE$_i$, NODE$_j$)=hd(i,j)
   =0 when Dist(NODE$_i$, NODE$_j$) hd(i,j)
   hd(i,j): number of difference in hierarchy layers between NODE$_i$ and NODE$_j$
   N: total number of related nodes for some communication
   NODE$_i$: ith node number (i = 1~ N)
   Dist(i,j): distance between NODE$_i$ and NODE$_j$
   $_n C_2$: n(n-1)/2

Ev value for Nodes C,E,F in Figure 5b becomes 0.33 from Equation 2.

5. A PROTOTYPE SYSTEM FOR EVALUATION OF ORGANIZATIONS FROM THE VIEWPOINT OF COMMUNICATION

A prototype system to evaluate organizational structure in emergency situations from the communication viewpoint was developed based on the indexes mentioned in the previous section using Java language on a PC.

5.1 System Architecture

The system is composed of the following four components.

Data input part: Input of data for physical structure and human related factors are managed.
Simulation management part: Evaluation indexes are calculated using simulation results.
Simulation part: Simulation results are calculated based on the communication model.
Interface part: This part is composed of three windows. Using these windows, simulation condition is controlled and calculated indexes are displayed to the user.

Figures 7, 8, and 9 show concrete interface windows for the prototype system. The function of each window is as follows.

(a) Window for simulation condition control (see Figure 7): Data selection and condition of simulation is controlled through this window. The window is composed of a-1 to a-5 parts. Output format is selected through a-1, a-2, a-3, a-4, and

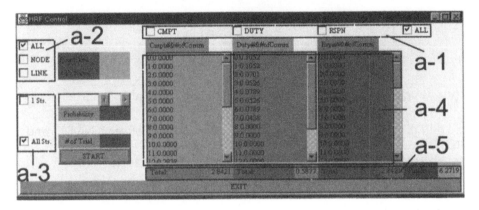

FIGURE 7 Window for simulation condition control.

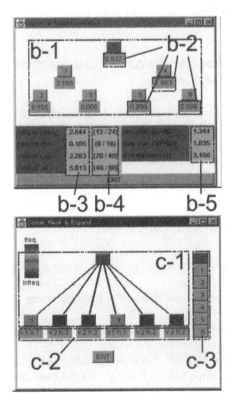

FIGURE 8 Window for evaluation result display.

FIGURE 9 Window for communication frequency display.

a-5, which show raw data of communication occurrence obtained from the simulation results.

(b) Window for evaluation result display (see Figure 8): Each value of the evaluation indexes is displayed in the window. In the figure, each part indicates the following data.

b-1: Organizational structure
b-2: Relative number of communication occurrences
b-3: Relative usage frequency for each human related factor data
b-4: Ratio between assigned human related factors and used ones
b-5: Index of horizontal extension and that of vertical extension

(c) Window for communication frequency display (see Figure 9).
Communication distribution between each node of the organization is displayed on the window. The top node is selected through c-3. c-1 shows communication distribution and c-2 shows horizontal and vertical distance.

5.2 Examples of Evaluation Results

If different organizational structures are given to the system, one can compare the structures from the viewpoint of communication and evaluate the merits of both structures. Here a hierarchical versus a flat structure are compared quantitatively from the communication viewpoint.

Organization 1: The organization is composed of three layers, and role division between upper layer and lower layer is determined explicitly.
Organization 2: The organization is composed of two layers, and human related factors are assigned to lower layers as much as possible.

Figure 10 shows concrete structures for each organization, and the same physical structure as in Figure 5 is used in this case. Figure 10 also indicates the evaluation indexes obtained from the developed prototype system for each organization. The following is the summary of the evaluation results.

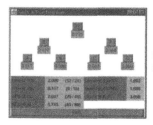

(a)
Organization 1

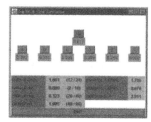

(b)
Organization 2

FIGURE 10 Evaluation results.

1. From the obtained values shown in Figure 10, communication is concentrated on both Node E and F in Organization 1. On the other hand, there is no communication bottleneck in Organization 2, and the desired flat distribution of communication occurrence is realized.

2. The communication frequency for other nodes is relatively low in Organization 2 as compared to Organization 1. On the index of extension, the value of Eh(Index of Horizontal Extension) in Organization 1 is a bit larger than that in Organization 2. On the value of Ev (Index of Vertical Extension), its value in Organization 1 is much larger than that of Organization 2. This means that Organization 1 solves problems via the communication between members of different layers. That is, collaboration between upper layers and lower layers is principal. However, in Organization 2 problem solving is conducted in a distributed style.

In this way, it is demonstrated that different organizations in an emergency can be evaluated quantitatively from the viewpoint of communication using the developed prototype system. Another usage of this system may be to find communication problems associated with a current system and to improve on them by changing human related factors data. This will also be useful for designing optimal organizational structure in emergency situations.

6. DEVELOPMENT OF GROUP COMMUNICATION SUPPORT SYSTEM IN EMERGENCY SITUATIONS

The communication model proposed in section 3 can be used for another application, a group communication support system for emergency situations. The evaluation system of organizational structure discussed in this article so far is basically for offline use to find communication problems and to improve organizational structure. On the other hand, group communication support systems in emergency situations are for online use and their objective is to give appropriate advice on communication in the confusion of an emergency situation. In the following sections, the issue of leveraging the communication model to aid in the design of group communication support systems is discussed.

6.1 Basic Idea of Group Communication Support System in Emergency Situations

As written in section 3, the communication model can predict which type of communication occurs and who the person to be contacted is when an emergency situation arises in an organization. Therefore the model can be used to build an advice system to tell who should be contacted for a given situation. We think it is very important to give advice on contact address in the confusion of emergency situations, as indicated by large scale disaster experiences.

In this application, it is assumed that the exact data of physical structure and human related factors are given to the system in advance. Although the difficulty of gathering these data in real time depends on each application, it seems that physical structure data is basically fixed for many plants, such as chemical or power plants. On the other hand, human related factors are basically decided in advance in each organization. However, the name of the person related to each factor changes from moment to moment because of factors such as rotation of plant management teams. Therefore, it is necessary to obtain data concerning organizational structure at the time of an incident in order for the advice system to work exactly.

We have designed a prototype of a group communication support system for emergency situations that gives advice on contact address in response to an emergency situation. Figure 11 shows an image of input and output from the prototype system. The input situation is expressed by the state of each sensor, and in this case, the data that "No. 3 sensor becomes high" are given to the system. Then the system gives an explanation of the current situation. In this case, the situation is that the flow of part "Comb" of plant "Pa" has become "H(igh)." In the next stage, the system indicates the action to be taken and the appropriate type of communication and contact address as shown in the figure. In this case, the action to be taken is to operate Actuator 5 down to decrease the flow of part "In-s" of plant "Pa." The type of communication and contact address for the example case are as follows: Communication generated by competence is to Person A, there is no communication generated by duty, communication generated by responsibility is to Person A, and communication generated by knowledge is to Person C. The parameters shown in the responsibility part determine what type of responsibility he or she has. In the example, it shows that Person A has the responsibility to keep the flow of "Comb" of plant "Pa" at the middle level. In the same way, the parameter in the knowledge part determines what type of knowledge he/she has. In the example, it shows that Person C knows "Comb" of plant "Pa" very well.

Input to the system:
  (sen, 3, H)

Explanation from the system:
  Sensor No.3, which indicates Flow of part "Comb" in Plant "Pa" becomes High
  (Pa, Comb, flow, H)

  Operate Actuator No.5 Down to decrease Flow of part "In-s" in Plant "Pa".
  (adj, 5, D)(Pa, In-s, flow, M)

Communication:
  Competence: Person A
  Duty: (No one)
  Responsibility: Person A
  Person A has a responsibility to guarantee (Pa, Comb, flow, Middle)
  Knowledge: Person C (Pa, Comb)

FIGURE 11 Image of the output used for a group communication support system.

6.2 Consideration of Work Flow and Exceptional Processing

The previously mentioned function can be easily realized based on the communication simulator, which was already developed as an evaluation system. However, it is strongly recommended that the following two functions be included in a group communication support system for emergency situations from the practical viewpoint in our investigation.

1. Integration of work flow system: The function of the workflow system presents the sequences of work steps to be taken in a work process. This system is widely used in offices, and it has become an indispensable function in groupware. Workflow should be added to the communication simulator in order to integrate both functions. Sequences of actions to be taken are set in the data of the TOS part of the communication simulator.

2. Support of dynamic substitute processing: The situation is sometimes observed in emergency situations that the person to be contacted is not at his or her chair when an emergency happens or the physical communication route is shut down because of some trouble caused by the emergency. In these cases, it is impossible to communicate with the appropriate person and for the processing to proceed for handling the emergency; it is thus necessary to support the persons in the organization who must cope with these problems. We call this "Support of dynamic substitute processing." Dynamic substitute processing is based on the rules provided in advance to decide the candidate for the substitute, and contact address and workflow are changed according to the candidate for the substitution.

Figure 12 shows the architecture of the whole group communication support system that includes the function of workflow and support of dynamic substitute processing. The first version of the prototype system was developed on a PC. Figure 13a shows an interface panel of input workflow data and rules to decide candi-

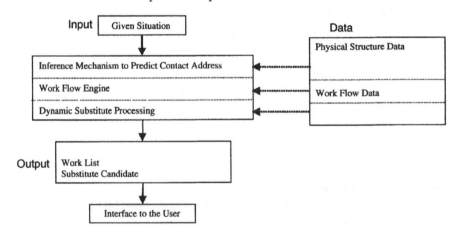

FIGURE 12 Architecture of group communication support system.

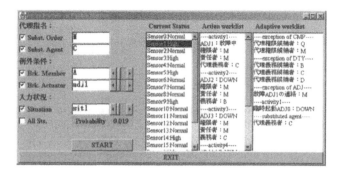

(a) Interface panel

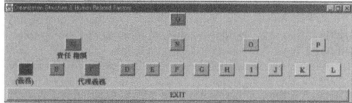

(b) Output panel

FIGURE 13 Prototype system for group communication support system.

dates for substitution. Figure 13b shows an organizational structure and the substitute person is displayed on this panel if the person to be contacted is absent or cannot be communicated with because of a physical accident. It is herein demonstrated that the communication model proposed in section 3 can be used to aid the design of a group communication support system for emergency situations, and also that workflow integration and support of dynamic substitute processing are possible within this design paradigm. The next step is to evaluate the group communication support system in a practical situation.

7. CONCLUSIONS

In this article, communication in emergency situations was analyzed in a hierarchical organization, and important factors were investigated from the viewpoint of communication. Then, a communication model that considers human related factors such as competence, duty, responsibility, and knowledge was proposed, and a prototype system to evaluate organizational structures during emergency situations was developed based on the communication model. As a result, it was demonstrated that the system can be used to quantitatively compare different organizations from the viewpoint of communication. Furthermore, the possibility of the design of group communication support systems based on the communication model was also discussed. We next plan to develop a group communication support system for emergency situations that includes workflow integration and support of exceptional processing.

REFERENCES

Conklin, J., & Begeman, M. L. (1988). gIBIS: A hypertext tool for exploratory policy discussion. *Proceedings of CSCW'88*, 140–152.

Ishii, H. (1990). Team work station: Towards a seamless shared workspace. *Proceedings of CSCW'90*, 13–26.

Kawabe, K., & Sakata, S, (1990). Distributed multiparty desktop conference system: MERMAID. *Proceedings of CSCW'90*, 27–38.

Kawata, Y. (1995). *Toshi dai saigai [Catastrophic urban disasters]*. Tokyo: Kinmiraisha.

Malone, T. W. (1987) Modeling coordination in organizations and markets. *Management Science*, 33(10), 1317–1332.

Malone, T. W., Grant, K. R., Lai, K. Y., Rao, R., & Rosenblitt, D. (1986) Semi-structured messages are surprisingly useful for computer-supported coordination. *Proceedings of CSCW'86*, 102–114.

Nakatani, M., & Nishida, S. (1992). Trouble communication model in a software development project. *IEICE Trans. Fundamentals*, 75-A(2), 196–206.

Nunamaker, J. F. (Ed.). (1989). Special issue on GDSS [Special issue]. *Decision Support Systems*, 5(2).

Okada, K., Ichikawa, Y., Jeong, G., Tanaka, S., & Matsushita, Y. (1995). Design and evaluation of MAJIC videoconference system. *Proceedings of INTERACT'95*, 289–294.

Winograd, T. A. (1988). Language perspective on the design of cooperative work. *Proceedings of CSCW'88*, 203–220.

INTERNATIONAL JOURNAL OF HUMAN–COMPUTER INTERACTION, *17*(1), 43–60
Copyright © 2004, Lawrence Erlbaum Associates, Inc.

InterActor: Speech-Driven Embodied Interactive Actor

Tomio Watanabe
Faculty of Computer Science and System Engineering
Okayama Prefectural University
CREST of Japan Science and Technology Corporation (JST)

Masashi Okubo
Faculty of Computer Science and System Engineering
Okayama Prefectural University

Mutsuhiro Nakashige
Graduate School of Engineering
The University of Tokyo

Ryusei Danbara
Graduate School of System Engineering
Okayama Prefectural University

A speech-driven embodied interactive actor called InterActor, with functions of both speaker and listener, was developed for activating human interaction and communication by generating expressive actions and motions coherently related to speech input. InterActor is the electronic media version of a physical interaction robot called InterRobot for robot-mediated communication support, which is free of the hardware restrictions for human interface of advanced graphical user interface based network communication. By using InterActor, the concept of speech-driven embodied interaction system is proposed for human interaction sharing by the entrainment between human speech and InterActor motions in remote communication. The prototype of the system is developed, and the sensory evaluation and behavioral analysis in human communication through InterActor demonstrates the effectiveness of the system. Actual applications of InterActor to human interface are also demonstrated. The system is a practical communication support system, which activates human interaction and communication on the basis of only speech input.

This work under our project E-COSMIC (Embodied Communication System for Mind Connection) has been supported by CREST of JST.

Requests for reprints should be sent to Tomio Watanabe, Faculty of Computer Science and System Engineering, Okayama Prefectural University, 111 Kuboki, Soja, Okayama 719–1197, Japan. E-mail: watanabe@cse.oka-pu.ac.jp

1. INTRODUCTION

In human face-to-face communication, both verbal and nonverbal communication such as nodding and body motions are rhythmically related and mutually synchronized between talkers (Condon & Sander, 1974; Kobayashi, Ishii, & Watanabe, 1992). This synchrony of embodied rhythms in communication, called *entrainment*, generates the sharing of embodiment in human interaction, which plays an important role in essential human interaction and communication. Hence, in remote non-face-to-face communication, the introduction of the entrainment mechanism to human interface is indispensable to the realization of essential human interaction and communication systems.

In conventional telecommunication systems such as a TV conference system, however, because the spatial relationships between talkers are cut off, it is difficult to share their embodiment in human interaction. By the immerse projection display technology, therefore, sharing the embodiment in interaction with a high quality of presence in virtual image space is examined (Cruz-Neira, Sandin, & DeFanti, 1993; Hirose, Ogi, Ishiwata, & Yamada, 1998). On the other hand, without setting this kind of spacious system, the sharing of embodiment by the entrainment would be realized by presenting a communication environment where the interaction between talkers is perceived, which would be expected to support human interaction and communication (Kishino, Miyasato, & Terashima, 1995; Morishima, 1999; Watanabe, 2001).

We already analyzed the entrainment between a speaker's speech and a listener nodding and blinking in face-to-face communication, made an interaction model and developed a speech-reactive system in which computer graphics representing human facial expressions simulated nodding and blinking in response to speech input. We demonstrated that the system can be effective for smooth speech input (Watanabe & Higuchi, 1991; Watanabe & Yuuki, 1989). The nodding reaction model was applied to not only the electronic media of computer graphics but also the physical media of robots, and the effectiveness of the model was confirmed (Yatsuka, Kawabata, & Kobayashi, 1997, 1998). We also proposed an embodied virtual communication system where the talker interactive behavior such as nodding, blinking, head and body motions, paralanguage, respiration, and so forth on the basis of his or her nonverbal and physiological information is freely controlled to clarify human interaction systematically in a virtual face-to-face communication environment. The importance of sharing mutual embodiment in communication was pointed out (Watanabe, & Okubo, 1999). Moreover, the speech-driven embodied interaction robot called InterRobot, shown in Figure 1, was developed, and the effectiveness was demonstrated in supporting human interaction and communication (Ogawa, & Watanabe, 2001; Watanabe, Okubo, & Ogawa, 2000)

In this article, focusing on the human interface of advanced GUI based communication, a speech-driven embodied interactive actor called InterActor with both functions of speaker and listener is developed for expanding the field of its application. InterActor is the electronic media version of InterRobot, which is free of hardware restrictions. By using InterActor, the concept of speech-driven embodied interaction system is proposed for supporting human interaction and communication with the sharing of embodiment by entrainment. The prototype of the system is developed,

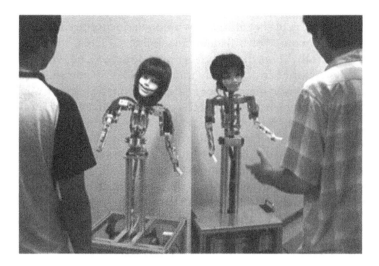

FIGURE 1 Remote communication using InterRobots.

and the sensory evaluation and behavioral analysis in human communication through InterActor demonstrates the effectiveness of the system. Finally, actual applications of InterActor in human interface are demonstrated.

2. EMBODIED INTERACTION

2.1. Concept

The concept of the embodied interaction system for supporting human interaction and communication is shown in Figure 2. The system consists of two InterActors that have both functions of speaker and listener on the basis of speech input. When Talker 1 speaks to InterActor 2, InterActor 2 responds to the utterance in an appro-

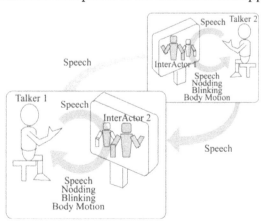

FIGURE 2 Concept of the embodied interaction system.

priate timing by means of its entire body motions, including nodding, blinking, and actions in the manner of a listener for Talker 1. Thus, the talker can talk smoothly and naturally. Then, the speech is transmitted via a network to the remote InterActor 1. InterActor 1 can effectively transmit Talker 1's message to Talker 2 by generating body motions in the manner of a speaker on the basis of a time series of the speech, and presenting both the speech and the entrained body motions simultaneously. Talker 2 in the role of a speaker this time achieves communication by transmitting his or her speech via InterActor 1 as listener and InterActor 2 as speaker to Talker 1 in the same way. In this way, two remote talkers enjoy conversation via InterActors. The only information transmitted and received by this system is speech. It is significant that it is a human that transmits and receives the information. InterActor just generates the entrained communicative motions and actions on the basis of human speech and supports the sharing of mutual embodiment in communication.

2.2. System Configuration

The outline of the system is shown in Figure 3. The system consists of a PC with 128 MB memory and a sound card, a microphone, and a loudspeaker as input and output devices. Microsoft DirectX 6.0 SDK is used for drawing InterActor with Microsoft Visual C++. The excursion file size is 300 KB. The system can be used on a general PC (CPU of the system is Pentium III-500MHz.). The degree of freedom of InterActor is 20, which is almost the same as InterRobot. InterActor expressive actions and motions are generated by binary control of each action (i.e., nodding; blinking; mouth movement; the rotation of neck, shoulder, elbow, and waist; and the bending of wrist, elbow, arm, neck, and waist). Human speech and binary control signals are recorded in HDD for representation of InterActor and interaction analysis.

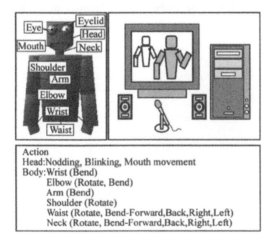

FIGURE 3 Outline of the system.

3. EMBODIED INTERACTION MODEL

3.1. Relation of Human Motions to Speech

Two participants faced each other and one participant talked to the other, each participant assuming the role of a speaker and a listener. The behavior and the speech were recorded on the same frames with a video editing system (SONY FXE-100) by two CCD cameras, one for the speaker and the other for the listener. The participants were 8 male students in pairs. The motions of head, both forearms, and body were measured by four magnetic sensors (Polhemus FASTRAK) put on the top of the talker head, both forearms and back. The parameter of nodding was defined as binary on or off according to whether the nodding response existed or not in the unit of 50 msec. The speech parameter was also defined in each 50 msec at a sampling rate of 11,025 Hz as the binary burst-pause of speech. Talkspurt or silence was discriminated by whether the speech power level was over or under a threshold value; this threshold was 12 dB plus the background noise level presented in the silence parts. The short silence duration before an unvoiced consonant is, however, caused by inertia of the vocal organ. Hence, the burst-pause of speech was generated with a hangover value of 150 msec where the hangover operation extended each talkspurt duration and converted silence duration less than or equal to 150 msec into talkspurt. Figure 4 shows an example of the analysis of the speech, the speaker motions, and listener motions. The relation between these parameters was evaluated using the cross-correlation function:

$$C(\tau) = \frac{\sum\limits_{i=1}^{n-\tau}\left\{x(i)-\mu_x\right\}\left\{y(i+\tau)-\mu_y\right\}}{\sqrt{\sum\limits_{i-1}^{n}\left\{x(i)-\mu_x\right\}^2}\sqrt{\sum\limits_{i=1}^{n}\left\{y(i)-\mu_y\right\}^2}} \tag{1}$$

where m_x and m_y are the mean value of x and y respectively. The analyzing time is 120 sec, $n = 24,000$ in 50 msec unit, in the region where the time lag τ is within 5 sec.

Figure 5 shows the typical results. In the correlations between the speaker's speech and the listener's motions, we can find the remarkable negative relation between the speech and the head motion that includes the essential action of regulator for the flow of conversation such as nodding. The correlations of the motions of arms and body to the listener own head motion were higher than those of the motions to the speaker speech. This demonstrates the importance of the relation of the listener head motion to the speaker speech motion.

In the correlations between speech and motions in the speaker himself, the correlation between the speech and the head motion was the highest, and the correlations of the motions of arms and body to the speaker own head motion were higher than those of the motions to the speech. These also demonstrate the importance of the relation with the head motion. As for the arm motion in the

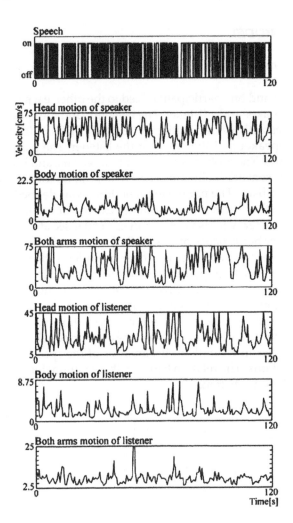

FIGURE 4 Time series of speaker's speech, speaker's motions and listener's motions.

speaker himself, the evaluation of relation between the burst-pause of speech and the binary data of arm motion in some thresholds is shown in Figure 6. It is demonstrated that the correlation between the speech and the arm motion in a high threshold (A-B) increases. These similar tendencies were shown in the other three pairs of talkers.

3.2. Interaction Model as Listener

From the analysis of the relation between the speaker speech and the listener motions in Figure 5, the importance of the relation with the listener head motion, which mainly included nodding to speech, was pointed out. Thus, as a listener interaction model, in addition to the nodding reaction model and the eye-blinking model that we have already proposed, a bodily reaction model of arms and body based on the nodding reaction model was introduced as follows. In the prediction

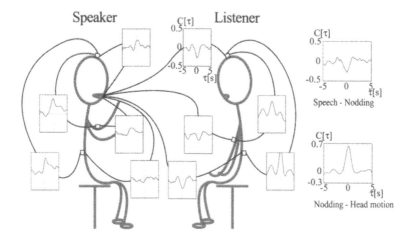

FIGURE 5 Relation among speaker's speech and his own motions and listener's motions.

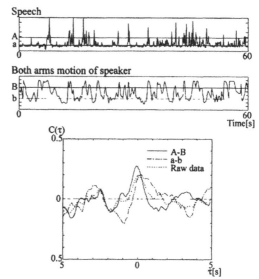

FIGURE 6 Evaluation of the relations between speech and both arms motion in speaker.

of nodding, the estimator was the Moving-Average (MA) model that estimated the nodding $y(i)$ as the weighted sum of the binary speech signal $x(i)$ in each 50 msec (Watanabe & Yuuki, 1989):

$$\hat{y}(i) = \sum_{j=1}^{J} a(j)x(i-j) \tag{2}$$

where $a(j)$ is the linear prediction coefficient.

The eye-blinks were generated with the timing of exponential intervals from each nodding onset in response to speech input, because the listener eye-blinks took place during nodding simultaneously and the eye-blinking intervals were re-garded as an exponential distribution (Watanabe, 1993). The bodily motions were related to the speech input by operating one of the 12 positions such as the 4 positions of neck, the 4 positions of waist, the 2 positions of right elbow, and the 2 positions of left elbow at the timing over the threshold that was lower than that of the prediction of nodding. In other words, for the generation of body motions of InterActor as listener, the time relations between nodding and the other motions were realized by varying the threshold values of nodding estimation. Figure 7a shows an example of InterActor action as listener.

3.3. Interaction Model as Speaker

From the analysis of the communication experiment in Figure 5, the speaker bodily interaction model based on the burst-pause of speech and the arm operation based on the power of speech were introduced to InterActor as speaker as follows. Speech input was in real time through a microphone or an audio file, and the output was from a loudspeaker. The mouth motion of InterActor was realized by the switching operation synchronized with the burst-pause of speech. The body actions of InterActor were also related to the speech input by operating both the neck and one of the other body actions as mentioned in the previous section at a timing over the threshold that was estimated by its own MA model of the burst-pause of speech to the head motion. Because speech and the motions of the arms were related at a relatively high threshold value, one of the arm actions in preset multiple patterns was selected for the operation of InterActor when the power of speech was over the threshold as shown in Figure 6. An example of InterActor action as speaker is shown in Figure 7b.

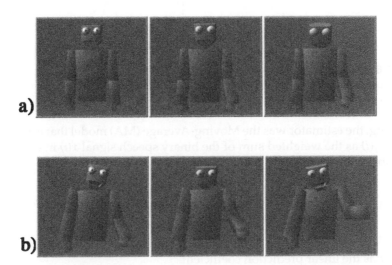

FIGURE 7 Example of speech driven InterActor actions. (a) Listener action. (b) Speaker action.

4. EVALUATION OF INTERACTOR SPEECH-DRIVEN MOTIONS

The mentioned embodied interaction models were applied to InterActor, and InterActor speech-driven expressive actions and motions were evaluated. The experiment was examined to make InterActor act for 150 sec as listener and speaker. Participants were 10 male students. Control signals for moving InterActor were used as InterActor motion data in which the motion was defined as binary on or off with a hangover value of 150 msec according to whether the motion acted or not. The speech was also defined in each 50 msec as the binary burst-pause of speech with a hangover value of 150 msec. The analyzing time was 120 sec in the center of 150 sec in the region in which the time lag τ was within 5 sec.

4.1. InterActor as Listener

Figure 8 shows an example of the analysis of human speech and the body motions of InterActor as listener, and cross-correlations between them. There is a significant negative relation without time lag between human speech and InterActor nodding, which denotes that InterActor responds at the timing equivalent to human nodding. The cross-correlations with the other body motions are also very similar to those in human face-to-face communication. It is clear that InterActor creates reactions by means of its entire body in close association between speech and body motions, including nodding. This demonstrates the effectiveness of InterActor as listener. Similar trends were also found for the other participants.

4.2. InterActor as Speaker

The result of evaluating the cross-correlations between the speech of human speaker and the body motions of InterActor as speaker is shown in Figure 9. There are significant positive relations between human speech and InterActor head, arms, and body motions, which are very similar to those in a human speaker. It is clear that the relations between the speech and the motions, centering on the motions of the head, are well expressed. This demonstrates the effectiveness of InterActor as speaker. These trends were found to be similar for the other participants.

5. INTERACTION BY USING INTERACTOR

By introducing both listener and speaker interaction models, InterActor with functions of both listener and speaker was realized. By using InterActor, a speech-driven embodied interaction system was developed. Sensory evaluation was examined for the system under two conditions of speaking to InterActor by one user and remote communication by two users. A microphone was used for speech input.

FIGURE 8 Time series of a speaker's speech and the motions of InterActor as listener, and C(τ) between them.

5.1. One User Interaction

A participant spoke to InterActor as listener on display by himself as shown in Figure 10. InterActor behaved as listener on the basis of the speech. The experiment was examined under the following three different modes and in the following order:

Mode (I): only one InterActor as listener (see Figure 11a).

Mode (II): two InterActors including the participant own InterActor as speaker in virtual face-to-face scene from the diagonal backward viewpoint of his own InterActor (see Figure 11b).

Mode (Ir): one InterActor as listener that generates communicative motions randomly in correspondence to the burst-pause of speech.

The experiment time was 150 sec in each mode and then 120 sec over switching modes freely.

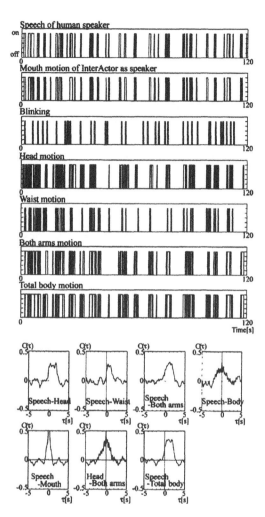

FIGURE 9 Time series of a speaker's speech and the motions of InterActor as speaker, and $C(\tau)$ between them.

FIGURE 10 Example of the 1 user's interaction scene.

FIGURE 11 Example of interaction mode (I) and (II).

The result of paired comparison is shown in Table 1. In the table, the Bradley-Terry model was assumed to evaluate the preference of modes in Table 1 quantitatively, defined as follows:

$$P_{ij} = \frac{\pi_1}{\pi_i + \pi_j}$$

$$\sum_{i=1}^{n} \pi_i = const.(= 30) \tag{3}$$

where π_i is the intensity of preference for mode i, and P_{ij} is the probability of judgment that i is better than j.

Table 1 shows π_i solved by the maximum likelihood method. The matching of the model was examined by the goodness-of-fit test:

$$\chi_0^2 = \sum\sum \frac{(X_{ij} - X_{1ij})^2}{X_{1ij}}$$
$$= 1.04 \tag{4}$$
$$\chi^2(1, 0.05) = 3.84 > \chi_0^2$$

Table 1 : Result of Questionnaire by Paired Comparison

	(I)	(II)	(Ir)	Total	π
(I)	—	6	9	15	17.75
(II)	4	—	6	10	8.34
(Ir)	1	4	—	5	3.91

Note. mode (I) = 1 InterActor; mode (II) = 2 InterActors; mode (Ir) = 1 InterActor (on-off sync.).

and the likelihood ratio test:

$$r = 2\sum\sum X_{ij} \times \log \frac{X_{ij}}{X_{1ij}}$$
$$= 1.08 \tag{5}$$
$$\chi^2(1, 0.05) = 3.84 > r$$

where X_{ij} is the number of judgment that i is better than j, and X_{1ij} is the expected value of X_{ij}.

These results indicate the validity of π_i and the matching of the model because the model was not rejected.

The sensory evaluation based on only speech communication is in Figure 12. Even mode (Ir) with communicative motions corresponding to the burst-pause of speech was evaluated as nearly "good" because human characteristics of nodding and blinking were introduced to InterActor in the same way. In comparison with the mode (Ir), mode (I) was evaluated highly in any item of preference, interaction, and enjoyment at the significance level of 1%. This demonstrates the effectiveness of InterActor as a listener.

5.2. Two Users Interaction

The experimental setup of remote communication using InterActors is shown in Figure 13. The situation of the communication experiment such as the correspondence of motions between talkers and InterActors was recorded through a video editor (SONY FXE-100) by two video cameras (SONY DV-CAM) as shown in Figure 14. Participants were 10 male students. Each participant talked in a separate room, and his speech was transmitted and received via 100 Mbps Ethernet. The experiment was examined with five pairs of 10 talkers on two different commu-

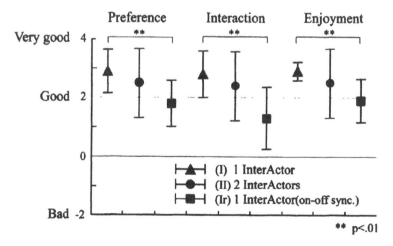

FIGURE 12 Result of sensory evaluation based on speech communication.

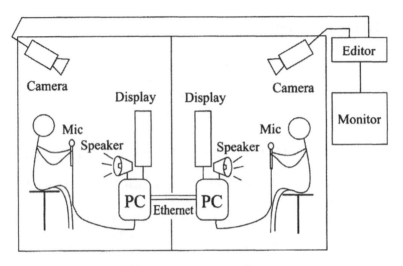

FIGURE 13 Experimental setup.

nication modes of only one InterActor in mode (I) and two InterActors including the talker own InterActor in mode (II). The experiment time was 150 sec in each mode and then 120 sec over switching modes freely. The result of sensory evaluation based on only speech communication is shown in Figure 15. It is found that both mode (I) and mode (II) are preferred in any item of preference, interaction, and enjoyment, which demonstrates the effectiveness of the system. As for enjoyment, mode (II) including the talker own InterActor was more highly evaluated than mode (I) at the significance level of 5%. Mode (II) was also selected by 8 out of 10 talkers by paired comparison of preference between (I) and (II). Figure 16 shows an example of virtual face-to-face communication scene of InterActors including themselves where two remote talkers can share the embodied interaction on the basis of speech.

FIGURE 14 Remote communication using the embodied interaction system.

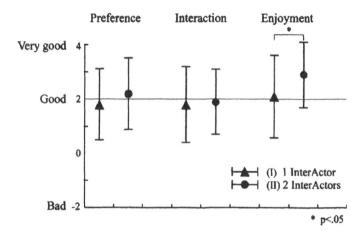

FIGURE 15 Result of sensory evaluation based on speech communication.

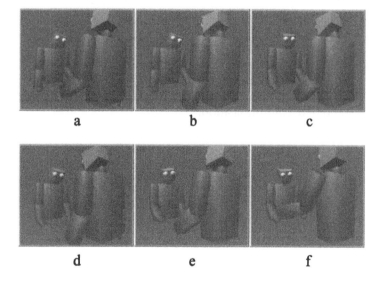

FIGURE 16 Example of virtual face-to-face communication scene.

6. APPLICATIONS

Some actual applications of InterActor to human interface are introduced. InterActor as shown in Figure 17 is commercialized under the name of InterCaster by which news and media contents are effectively and cordially transmitted in a visual way of broadcasting with FTTH (Fiber To The Home) or

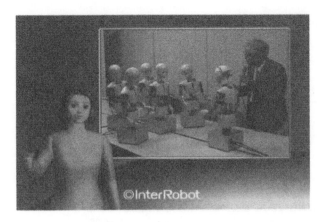

FIGURE 17 InterCaster.

in a commercial program. Figure 18 shows the speech-driven embodied group-entrained communication system called SAKURA. SAKURA activates group communication in which InterActors are entrained to one another as a teacher and some students in the same virtual classroom. By using SAKURA, talkers can communicate with the sense of unity through the entrained InterActors by only speech input via network. Figure 19 shows a physical version of SAKURA with InterRobots and InterActor, which is exhibited in the National Museum of Emerging Science and Innovation where visitors can enjoy a dynamic experience of embodied communication. They perceive the effects of the group-entrained communication environment intuitively and recognize the importance of embodied communication.

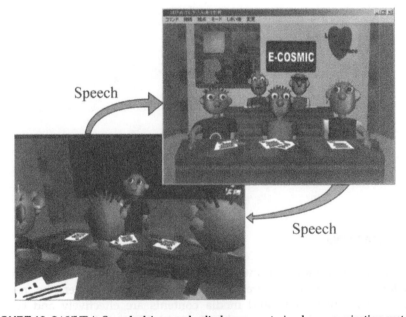

FIGURE 18 SAKURA: Speech-driven embodied group-entrained communication system.

FIGURE 19 Physical version of SAKURA with InterRobots and InterActor in National Museum of Emerging Science and Innovation.

7. CONCLUSIONS

In this article, a speech-driven embodied interactive actor called InterActor with functions of both speaker and listener was developed on the basis of nonverbal entrainment such as nodding and body actions and motions coherently related to speech in face-to-face communication. By using InterActor, the concept of a speech-driven embodied interaction system was proposed for supporting human interaction and communication with the sharing of embodiment. The prototype of the system was developed, and the effectiveness of the system was demonstrated. Some actual applications of InterActor to human interface were also demonstrated.

The system is a robust and practical communication support system, which activates human interaction and communication on the basis of speech input. InterActor and InterRobot technology, called iRT, would be expected to form the foundation of embodied communication technology as well as the methodology for the analysis and understanding of human interaction and communication, and to develop a new embodied communication industry for supporting essential human communication.

REFERENCES

Condon, W. S., & Sander L. W. (1974). Neonate movement is synchronized with adult speech: Interactional participation and language acquisition. *Science, 183,* 99–101.

Cruz-Neira, C., Sandin, D., & DeFanti, T. (1993). Surround-screen projection-based virtual reality: The design and implementation of the CAVE. *Proceedings of SIGGRAPH 3,* 135–142.

Hirose, M., Ogi, T., Ishiwata, S., & Yamada, T. (1998). Development and evaluation of immersive multiscreen display CABIN. *IEICE Transactions on Information and Systems, J81-D-II, 5,* 888–896.

Kishino, F., Miyasato, T., & Terashima, N. (1995). Virtual Space teleconferencing-communication with realistic sensations. *Proceedings of 4th IEEE International Workshop on Robot and Human Communication (RO-MAN 5),* 205–210.

Kobayashi, N., Ishii, T., & Watanabe, T. (1992). Quantitative evaluation of infant behavior and mother–infant interaction. *Early Development and Parenting, 1,* 23–31.

Morishima, S. (1999). Multiple points face-to-face communication in cyberspace using multi-modal agent. *HCI Communication, Cooperation, and Application Design, 2,* 177–181.

Ogawa, H., & Watanabe, T. (2001). InterRobot: Speech driven embodied interaction robot. *Advanced Robotics, 15, 3,* 371-377.

Watanabe, T. (1993). Voice-responsive eye-blinking feedback for improved human-to-machine speech input. *Advances in Human Factors/Ergonomics, 19B,* 1091–1096.

Watanabe, T. (2001). E-COSMIC: Embodied communication system for mind connection. *Usability Evaluation and Interface Design, 1,* 253–257.

Watanabe, T., & Higuchi, A. (1991). Facial expression graphics feedback for improving the smoothness of human speech input to computers. *Advances in Human Factors/Ergonomics, 18A,* 491–497.

Watanabe, T., & Okubo, M. (1999). An embodied virtual communication system for human interaction sharing. *Proceedings of 1999 IEEE International Conference on Systems, Man, and Cybernetics (SMC 9),* 1060-1065.

Watanabe, T., Okubo, M., & Ogawa, H. (2000). An embodied interaction robots system based on speech. *Journal of Robotics and Mechatronics, 12*(2), 126–134.

Watanabe, T., & Yuuki, N. (1989). A voice reaction system with a visualized response equivalent to nodding. *Advances in Human Factors / Ergonomics, 12A,* 396–403.

Yatsuka, K., Kawabata, K., & Kobayashi, H. (1997). A robot listener for fluent verbal communication. *IEEE RO-MAN 7,* 408–411.

Yatsuka, K., Kawabata, K., & Kobayashi, H. (1998). A study on psychological effects of human-like interface. *IEEE RO-MAN 8,* 89–93.

INTERNATIONAL JOURNAL OF HUMAN–COMPUTER INTERACTION, 17(1), 61–70

Remote Infrared Audible Signage System

Takuro Hatakeyama
Faculty of Rehabilitation, Seijoh University

Fumio Hagiwara
Hajime Koike
Keiji Ito
Hirohiko Ohkubo
Mitsubishi Precision Co. Ltd.

C. Ward Bond
Talking Signs, Inc.

Masao Kasuga
Faculty of Engineering, Utsunomiya University

When walking along streets or inside buildings, it is important for people with visual disabilities to acquire environmental information in order to update their mental map for accurate orientation as well as to ensure safe mobility. Various devices have been developed to acquire this information, but many problems remain unresolved. To overcome these difficulties, the authors describe two new additions to a Remote Infrared Audible Signage System (Talking Signs®) for use by people with visual disabilities that they can use not only in public places but also in the personal environment of their daily life. These efforts are currently taking place through a joint Japanese–U.S. company collaboration.

1. INTRODUCTION

For over 10 years the words "barrier free" have increased in popularity in Japan. However, there still exist many barriers to block disabled people from joining the broader social community. For people with visual impairments, orientation and

Appreciation is given for the kind cooperation, assistance, precious opinions and suggestions offered by Dr. William F. Crandall, Jr. of The Smith-Kettlewell Eye Research Institute, Masaaki Shiraishi of the Productive Aging Research Institute, Association for Technical Aids of Japan, Keio Plaza Hotel, and the Information-technology Promotion Agency of Japan.

Requests for reprints should be sent to Takuro Hatakeyama, Faculty of Rehabilitation, Seijoh University, 2–172, Fukino-dai, Tokai City, Aichi Prefecture 476–8588, Japan. E-mail: hatakeyama@seijoh-u.ac.jp

mobility based on a mental map is the most important component of successful travel to a destination. Several devices exist to assist these travelers, but these systems also have some recognized problems. Out of this need, we developed and deployed a device that efficiently and conveniently provides basic orientation and mobility information personally to a person with visual impairment. It can be used by itself or in conjunction with other assistive devices. We have been cooperating with Talking Signs Inc. (TSI, Baton Rough, LA, USA) since early in the development process. Talking Signs® is a registered trademark of TSI. Numerous evaluations of the usefulness of Talking Signs by blind travelers have been conducted by researchers at The Smith-Kettlewell Eye Research Institute in San Francisco and the Department of Geography, University of California, Santa Barbara. For example, studies are available for street crossing information (Crandall, Bentzen, Myers, & Brabyn, 1999), emergency exit notification (Crandall et al., 1999) and transportation applications (Crandall, Bentzen, Myers, & Brabyn, 2001; Marston & Golledge, 2000, in press).

2. THE PROBLEMS TO BE SOLVED

Some assistive systems for pedestrians (Mitabiya, 1995), like tactile block (tactile tiles), audible signals, and position indicator systems with voice messages via loudspeaker, are intended to provide the direction or information from environment facilities. In Japan, the tactile block is utilized at public facilities and intersections, but the design of the system has only recently been standardized and the cost for installing blocks is rather expensive. In addition, it is not possible to code the meaning of each possible destination to a particular kind of tactile marking.

Audible signals with melody, chirp, or speech sound from loudspeakers are used at many intersections and are popular (Bentzen & Tabor, 1998). However, there are some problems: Often it is not clear from which corner the sound is being sent; some pedestrian sometimes lose their proper orientation when crossing; and audible signals are noisy for shops and residents living near these intersections. Recently, a new audible system that aims at improving the crossing orientation problem has been developed. However, excessive noise pollution from such a loudspeaker system still remains a problem.

Another system uses micropower FM radio waves that are constantly transmitted. When people with an FM receiver near the transmitter, they receive a voice message (Bouoka, Sagara, & Akazawa, 1996). However, due to the nondirectional characteristics of radio signals, it is not possible for a person with visual impairment to get the needed directional information. In addition, there is the problem of cross talk between transmitters (Aono, Hatakeyama, & Tanaka, 1998). Recently, significant developments in the use of Global Positioning System (GPS) technology for automotive, aircraft, maritime, and even personal navigation (Loomis, Golledge, & Klatzky, 2001) have been deployed. However, for the purpose of blind navigation, certain important and inherit limitations in GPS technology are unlikely to be overcome. For example, there remain significant errors in GPS position accuracy, an inability for GPS to provide directional information, inoperability of

GPS within the dense urban high-rise environment and the inability of GPS to receive and process satellite information inside buildings.

We now describe a new audible signage system (Hatakeyama et al., 2001) for people with visual impairments that provides the required information personally and simply. It unambiguously gives "on-demand" directional and labeling information and solves the problem of noise pollution. In addition, this system works well both inside and outside of buildings.

3. SYSTEM DEVELOPMENT

Talking Signs technology was initially developed to privately provide direction and labeling information to people with visual impairments. It has also been determined to be useful in assisting individuals with other print-reading disabilities (Crandall et al., 1999). It is comprised of infrared transmitters that convey speech messages to small, handheld receivers. Infrared transmission is directional. This means that when people with visual impairments pick up a message, they can also tell the direction from whence it is coming; the message is coming from the direction in which they are pointing when they hear the message most clearly.

3.1. Design Policy

Design policies are as follows:

1. It is an assistive device, not replacing the white long cane.
2. It is not a safety device, although it reduces the need for concentration on a mental map.
3. It provides directional information.
4. It provides information to the user personally.
5. Information can be easily changed.
6. It is easy to learn to use the system.
7. It is easy to operate.
8. It is to be used by all people who have print-reading disabilities—including aged people—thus universal use is facilitated.

System use is shown in Figure 1.

3.2. Configuration

The system consists of transmitters and a handheld receiver (see Figure 2). The infrared transmitter has a voice recording module, FM modulation module, infrared transmitter module, and so on. The infrared receiver has an infrared sensor, demodulator, speaker, and so on. Transmitters are installed at entrances and exits of buildings, public telephones, elevators, toilets, and so on. That is, transmitters are installed as labels every place where information is required. Users carry the

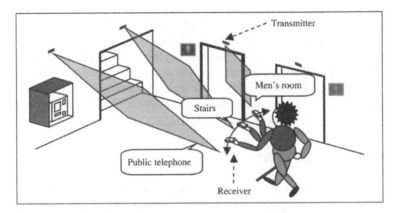

FIGURE 1 Utilization image of Talking Signs.

handheld receiver as they travel through the environment. Communication range can be adjusted from about 5cm to 30m.

3.3. Functions

The transmitter hosts the recorded voice message that is sent on an infrared (IR) beam by way of a frequency modulated (FM) subcarrier. When the user scans the receiver in the direction of the transmitter, the voice will be heard from the speaker within the handheld receiver. When the detector in the receiver faces the transmitter most directly, the voice is heard most clearly.

If the user moves the receiver off-axis to the transmitter, the voice will be heard with noise. If the angle of the detector is too great, then the voice will not be heard at all and the receiver will be silent. In this way, the user can easily and efficiently identify the correct direction to the transmitter (i.e., the direction to the information). The aperture angle of the receiving window is about 40° to 60°, which means that users can easily find any transmitters in the environment. The shape and dimensions of transmitters and receivers are shown in Figure 3.

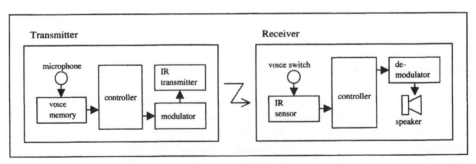

FIGURE 2 Block diagram of Talking Signs.

3.4. Features

The voice switch is designed to sit on the top center of the receiver case; right-handed and left-handed people can use it with equal ease. Also, it has been determined that this location is easier for wrist motion when scanning the receiver, particularly for people who are older and may have limited wrist flexibility. This switch position is also better for people wearing gloves. There is an earphone jack for people who prefer privacy or in extremely noisy environments. This jack can also be used for inductive loop for hearing aid compatibility. A three-position volume switch accommodates people with different hearing capabilities and people using the receiver in different noise conditions. The system is also effective for people with dyslexia in various ways: Those who may not easily understand written labels or directions may more easily understand the speech messages. Those who may have difficulty understanding directions or instructions given in speech may benefit from this system because the messages are constantly being repeated. Because this system labels the environment, it is useful for people who cannot read and understand a map properly or cannot maintain a mental map to their destination.

4. EXTENDED APPLICATIONS

In Japan, the Talking Signs system is installed in many facilities. There are many extended applications of which two typical examples—the Personal Locator System and the Network System—are described in some detail next.

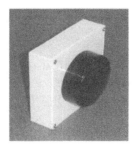

Transmitter
Head: 80 mm diameter

Receiver
W55 x H115 x D22 mm, 130g

FIGURE 3 Shape and dimensions of Talking Signs.

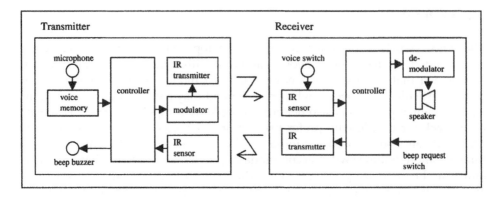

FIGURE 4 Block diagram of Personal Locator system.

4.1. Personal Locator System

Based on the original Talking Signs system, the new and important feature of the Personal Locator System (see Figure 4) is the true portability of the transmitter:

1. The transmitter is small and lightweight.
2. The transmitter is battery powered and self-contained.
3. The transmitter emits a "beep" sound when the receiver is nearby.
4. The transmitter has a "sleep mode" of operation. Therefore, it uses a very small amount of power when it is not being searched for (activated) by the receiver.[1]

The shape and dimensions for the units are shown in Figure 5.

Evaluation tests (see Table 1) of the Personal Locator System were held at the Keio Plaza hotel in Shinjuku, Tokyo and within the Shinkansen (bullet train) cars. Figure 6 shows the testing situation. Feedback from participants of the evaluation included the following comments:

1. The beep tone should be lower in frequency than is currently used.
2. The system will be more useful if coordinated with elevators.
3. It is better to set the transmitter at the pillow position of the seat in trains instead of on the seat cushion.

---

[1]For the transmitter "beep" and "wakeup" feature, the receiver itself sends a coded IR signal to the transmitter, prompting emission of either (a) a "beep" if the auxiliary pushbutton on the receiver is depressed or (b) the IR message if the main pushbutton on the receiver is depressed. Thus, for the Personal Locator System, communication is two-way.

Transmitter
W110 x H70 x D15 mm,
110g

Receiver with beep request switch
W55 x H115 x D22 mm, 130g

FIGURE 5 Shape and dimensions of Personal Locator system.

Table 1: Evaluation Test Results: Percentage of Effectiveness of Personal
Locator System

Items	Questionnaire	Hotel Room	Hotel Elevator	Bullet Train
Effectiveness	Is it useful?	90	85	100
	Is beep tone effective for some distance?	81	—	69
	Did you confirm that you arrived at target?	85	95	94
Relief	Is it relief to walk around with this system?	86	86	93

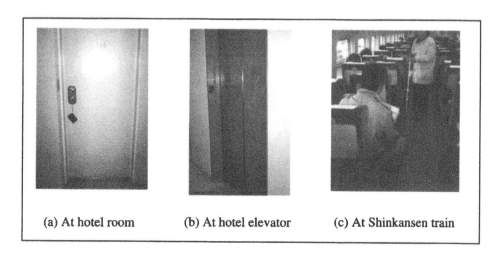

(a) At hotel room   (b) At hotel elevator   (c) At Shinkansen train

FIGURE 6 Evaluation test scene of Personal Locator system.

4.2. Network System

Based on the original Talking Signs system, the new and important feature of the Network System (see Figure 7) is the provision of real-time information:

1. Information is provided in real-time, on time, and at any time that it is required.
2. Operation is simple and easy to learn by the transit staff.
3. Future expansion of the system can be easily accomplished (good scalability).

The following functions are implemented in system software:

1. IP addresses management.
2. Voice message management.
3. Scheduled time database management.
4. Real-time information delivery.

The block diagram of the Network System is shown in Figure 8. Evaluation test of the Network System by people with visual impairments as well as aged people and responses by the pseudo administrator of the system (to a questionnaire) are shown in Table 2.

5. FUTURE ACTIVITIES

Infrared voice message from the bus destination signs will be tested in Japan in the near future. This system, currently deployed in several cities in the United States, informs awaiting passengers of the name and destination of the bus when the bus is 30 m away from the bus stop. Plans are underway to reduce the size of the receiver and to develop a model that is "hands-free." New work has begun on combining the directional attributes of IR with a wireless RF technol-

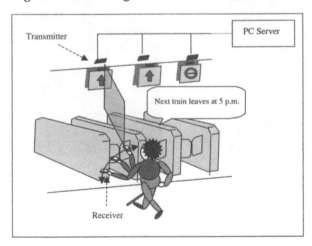

FIGURE 7 Utilization image at the ticket gate.

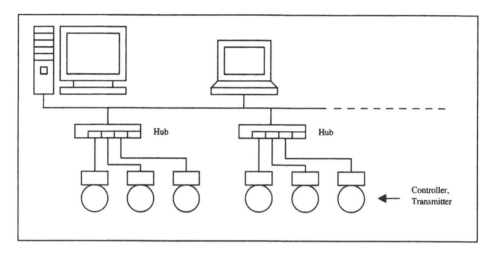

FIGURE 8 Block diagram of Network system.

Table 2: Evaluation Test Results: Percentage of Effectiveness of Network System

Evaluator	Items	Excellent	Very Good	Good	Fair	Poor
User	Effectiveness					
	Visually impaired	87.5	12.5	0	0	0
	Aged	85.0	15.0	0	0	0
	Relief					
	Visually impaired	83.3	16.7	0	0	0
	Aged	70.0	30.0	0	0	0
	Operability					
	Visually impaired	66.7	29.2	4.1	0	0
	Aged	85.0	10.0	5.0	0	0
System administrator	Operability	0	20.0	80.0	0	0
	Workload	0	40.0	20.0	40.0	0

ogy such as Blue-tooth or Wi-Fi (802.11X). Employing this hybrid approach, the user will be given on-demand control over the type and amount of information received, effectively allowing the user to "point and click" information in the environment.

To ensure that signage information is available to people who have print-reading disabilities such as blindness and mental retardation, it will be important to set an international standard so that a receiver and a user technique that works in one country will work in all other countries. If, like graphical signs, the system is installed everywhere, information will be seamless and can be universally used by ordinary people as well as people with disabilities. Such a design means universal design, and design for all.

6. CONCLUSIONS

A new infrared audible signage system for people with visual impairments and other print-reading disabilities has been developed that privately provides the user with accurate labeling and directional information to the destination. Enhancements described herein extend the usefulness of the basic Talking Signs system by providing portability of the system for personal use in, for example, locating one's hotel room or in locating one's train seat. Also discussed is a variant that provides real-time information to the blind traveler in the transportation terminal setting.

REFERENCES

Aono, M., Hatakeyama, T., & Tanaka, S. (1998, August). Survey on voice guidance system for people with visual impairments. *Proceedings of 13th Rehabilitation Engineering Conference,* 441–446.

Bentzen, B., & Tabor, L. (1998). *Accessible pedestrian signals* (U.S. Access Board, Contract No. PD–97–0772). Retrieved January 27, 2004 from http://www.access-board.gov/research&training/pedsignals/pedestrian.htm

Bouoka, M., Sagara, J., & Akazawa, Y. (1996, July). Voice guidance system using micro power radio wave. *Proceedings of 11th Rehabilitation Engineering Conference,* 237–288.

Crandall, W., Bentzen, B., Myers, L., & Brabyn, J. (1999, December). *Three studies for the National Institute on Disability and Rehabilitation research evaluating the effectiveness of remote infrared audible signs in solving current and emerging access problems of people who have print reading disabilities: Street crossing information, transit facility accessibility for people with developmental disabilities, and emergency egress information in buildings.* Retrieved January 27, 2004 from http://www.ski.org/Rehab/WCrandall/prolog.html

Crandall, W., Bentzen, B., Myers, L., & Brabyn, J. (2001). New orientation and accessibility option for people with visual impairments: Transportation applications for remote infrared audible signage. *Clinical and Experimental Optometry, 84*(3), 120–131.

Hatakeyama, T., Hagiwara, F., Koike, H., Ito, K., Ohkubo, H., Bond, W., et al. (2001). Remote infrared audible signage system. *Journal of Human Interface, 3*(3), 43–50.

Loomis, J., Golledge, R., & Klatzky, R. (2001). GPS-based navigation systems for the visually impaired. In W. Barfield & T. Caudell (Eds.), *Fundamentals of wearable computers and augmented reality* (pp. 429–446). Mahwah, NJ: Lawrence Erlbaum Associates, Inc.

Marston, J., & Golledge, R. (2000). *Towards an accessible city: Removing functional barriers for the blind and vision impaired: A case for auditory signs.* University of California Transportation Center. Retrieved January 27, 2004 from http://www.uctc.net/papers/423.pdf

Marston, J., & Golledge, R. (in press). The hidden demand for activity participation and travel by people who are visually impaired or blind. *Journal of Visual Impairment and Blindness.*

Mitabiya, H. (1995, August). Discussion on augmentative system for people with visual impairments. *Proceedings of 10th Japanese Conference on Advancement of Rehabilitation Technology, Rehabilitation Engineers Society of Japan,* 281–282.

INTERNATIONAL JOURNAL OF HUMAN–COMPUTER INTERACTION, 17(1), 71–88

Design of Interactive Skill-Transfer Agent From a Viewpoint of Ecological Psychology

Takayuki Shiose
Graduate School of Informatics, Kyoto University
ATR Intelligent Robotics and Communication Lab.

Tetsuo Sawaragi
Graduate School of Engineering
Kyoto University

Akira Nakajima
Hideshi Ishihara
OMRON Corporation

This article focuses on the design of an interactive skill-transfer agent for parameter tuning of an image sensor used to distinguish inferior goods from regular goods in a production line. The authors analyze the difficulty of transferring skills from a viewpoint of ecological psychology that takes into account the reciprocal relations between the participant and the environment. This article introduces an agent-based interactive skill-transferring system that stretches the meaning of this reciprocity as an interaction between an instructor and a successor. In concrete terms, an interactive agent system is proposed using an interactive learning classifier system with facial icons to enhance the human user's trust in the agent. The experimental results demonstrate the effectiveness of this system in transferring a human expert's skills.

1. INTRODUCTION

The purpose of this article is to develop an interactive skill-transfer agent. Recently, in Japanese industries, due to the prevailing advancement of automated systems, allowing next-generation workers to inherit skills from aged, experienced practitioners has become an urgent issue. That is, automation has cloaked the practitioners' proficient skills in a black box, and opportunities for human-to-human succession of these skills are becoming less frequent.

Thanks to Yasuhiro Nikaido for his invaluable help with experimental design and exciting discussions.
Requests for reprints should be sent to Takayuki Shiose, Graduate School of Informatics, Kyoto University, Yoshida-Honmachi, Sakyo-ku, Kyoto, 606–8501 Japan. E-mail: shiose@i.kyoto-u.ac.jp

Despite the seriousness of this situation, systemization enabling and promoting such skill transfer has not been attempted so far. This is because most of the skills belong to a human and are thus difficult to share and reuse. Moreover, as compared with other engineering issues such as maintenance and automation, it is difficult to obtain explicit effects of improvements on production efficiency and financial return, so the task of skill transfer has been left only to individual, voluntary efforts.

In this study, we focus on skilled workers engaged in tuning an inspection image sensor that can automatically distinguish inferior goods on production lines. This is a very difficult task for apprentices or probationers, and they must perform this task frequently (i.e., not only when an image sensor is first introduced at the factory site, but also whenever the objects or goods of inspection are changed). At present, the service engineer of the sensor vendor is called on demand whenever tuning is needed, and this task is dependent on new skills. The know-how related to such skills is difficult to describe, which creates a bottleneck for skill succession and transfer mediated by Information Technology (IT) such as artificial intelligence. Currently, the only method of transfer is through human-to-human education similar to an apprenticeship and on the job training (Rogoff, 1990).

In this article, to develop an IT-based system supporting such education and training, we propose an apprenticeship system mediated by an interactive agent. First, we classify some difficulties of skill-transfer in the practical field and insist that it needs the viewpoint of ecological psychology as a logical consequence. Second, we analyze the difficulty of transferring skills from a viewpoint of ecological psychology that takes into account the reciprocal relations between the participant and environment, and stress that it is important to pay attention to the interaction between the instructor and the successor. Next, we demonstrate our interactive agent system implemented by an interactive learning classifier system (ILCS) based on such characteristics. We then present the results of experiments using our system. Finally, we discuss the effects of skill transfer and its contribution to reducing the time costs incurred in the skill transfer from experts to apprentices by summarizing opinions we obtained from actual users who interacted with the agent.

2. TARGETS FOR SKILL-TRANSFER STUDY

2.1. Human Skills in the Process of Tuning Parameters

In this study, we focus on skilled workers engaged in tuning an image sensor that automatically distinguishes inferior goods in the streams of production lines (see Figure 1). When such a sensor is introduced into a factory, it must be tuned appropriately so that it functions as desired. However, this is a complex task because there is a variety of inspected targets and several environmental factors to be considered, such as lighting conditions and the speed of the lined streams.

More concretely, the main part of this task is to find a sequence of operations (e.g., to specify a particular location or to change the size of an inspection frame) that enables the sensor to force inferior parts out of the entire image area so that the inspection frame can include only the essential image features that distinguish the

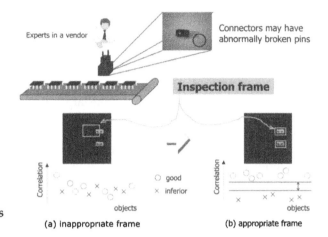

FIGURE 1 Skills in the process of tuning parameters.

(a) inappropriate frame

(b) appropriate frame

inferior from the others. Once this frame is determined and a sensor is appropriately tuned, it can automatically distinguish inferior goods. Although the skills needed to perform this tuning are very difficult for apprentices or probationers to perform, skill transfer from skilled workers to probationers takes place within a system of apprenticeship.

2.2. End Condition for Tuning Parameters

Even an expert cannot decide the form of the desired inspection frame until the expert encounters the inspection target in practice. The expert must decide on a tentative form of the inspection frame and check how accurately the image sensor can distinguish inferior goods from the production line using the provisional frame. The validity of the inspection frame is calculated by the following process, and Figure 2 shows a conceptual image of the tuning process.

First, a correlation value between the image of the target product and the image of a regular product is calculated in the range of the provisional inspection frame

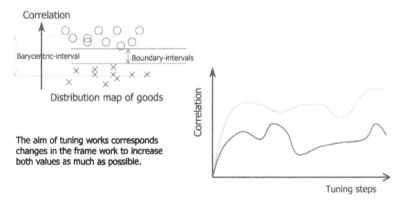

FIGURE 2 End conditions: Definition of barycentric and boundary-intervals.

that the user has temporarily chosen for the inspection of all objects registered in the process of tuning. Second, regular goods and inferior goods are marked differently for convenience. If the target product is categorized with the regular goods (we put a circle mark on the regular goods in Figure 2), the correlation value should be rated highly. On the other hand, the correlation value should achieve a low score when there is something wrong with the target product (we put a cross mark on the inferior goods in Figure 2). Third, the provisional inspection frame is validated in a visualized graph such as that in Figure 2. If the inspection frame is set up properly, the correlation values for the regular goods will show a higher value and those for the inferior goods will show a lower value. Here, to clarify the progression of the tuning processes, two indexes are introduced: the barycentric interval and the boundary interval. The former is defined as the interval between the barycentric coordinates of both groups, and the latter is defined as the interval between the lowest correlation value of the regular production group and the highest correlation value of the inferior goods group.

Tuning can be interpreted as a process of finding a sequence of operations that increases the values of both the barycentric interval and the boundary interval as much as possible.

2.3. Difficulties in Skill-Transfer

Generally speaking, difficulties in skill-transfer occur during the expert's externalization of skills and the apprentice's embodiment of the externalized skills. Many approaches for transferring expertise have been studied but have failed to obtain satisfactory results, despite using information technology that challenged the former difficulty (Boring & Harper, 1948; Clancey, 1983; Davis, 1979; Shortliffe, 1974, 1976). With regard to the latter difficulty, Ueno and Furukawa (1999) and Norris (1993) not only succeeded in quantifying advanced skills in sports and music to extract implicit knowledge from an ergonomic viewpoint, but also pointed out that the compulsion of such quantified skills may import bad habits to the successor without considering the individual differences of body and learning pace.

Here, the reason why an apprenticeship system between a human instructor and human successor practically succeeds in transferring the expert's skill in our focus. The most important difference from common skill transfer mediated by information technology is the hypothesis with regard to the relationship between an expert and an apprentice. First, the expert is requested to externalize his or her own proficient skills by him or herself as usual, whereas the apprentice must discover the nuances of the skill actively by watching the expert in the apprenticeship system. Second, such discovered skills are not easily grasped by the apprentice, so that the apprentice needs the embodiment process of such skills.

We do not hypothesize that our system can transfer an expert's skills in their entirety to an apprentice, because skills contain both verbally describable and tacit features that are difficult to encode. We believe that it is possible to encode the verbally describable features of skills by developing an appropriate learning algorithm, but for the tacit features, an approach of isolating and extracting skills from a human and storing them in a machine is quite limited. Rather, we must develop another approach that can provide humans with the opportunities to exert their potential tacit skills and can let apprentices self-discover those skills.

In the next section, these approaches are explained and implemented from the viewpoint of ecological psychology.

3. SKILL-TRANSFER FROM A VIEWPOINT OF ECOLOGICAL PSYCHOLOGY

3.1. Viewpoint of Ecological Psychology

In ecological psychology, "to act" is regarded not as something static that is separated from the participant and the environment, but as something dynamic that emerges from interdependent relations between the participant and the environment. Brunswik (1956) was one of the founders of such ecological psychology. He proposed a lens model wherein relations between the judgment participant and the environment (distal) were explained by a symmetrical structure, placing cues (proximal) for the judgment at the center. Figure 3 shows an overview of the lens model, which is called a *double system design* in that it analyzes cues from two directions for the judgment and the environment.

An especially important point is that Brunswik distinguished "cue utilization validity" from "ecological validity." The former indicates the latent validity of how accurately cues specify the ecological criteria, and the latter indicates the individual validity of how efficiently cues reflect the acting participant's cognitive judgment (Cooksey, 1996, puts psychological meanings and details of the lens model together). As the logical conclusion of this distinction, the difficulties of skill-transfer are explained by introducing the validity, "cue utilization validity," and are classified into the following two processes: the expert's difficulty in externalizing his or

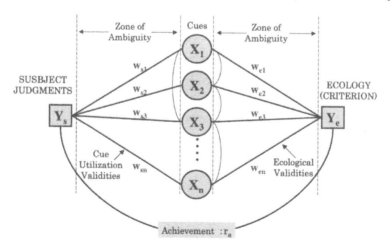

FIGURE 3 A concept of lens model.

her personal cues and the apprentice's difficulty in embodying such acquired cues. The latter difficulty particularly emphasizes that we cannot adopt the existing simple skill-transferring approaches to force the expert's skills to the apprentice or other target.

3.2. Skill-Transfer as an Interactive Activity

In the practice of an apprenticeship system composed of dyadic interaction between an expert and apprentices, the expert does not explicitly teach his or her skills to apprentices. This feature is called *positive concealment*, which encourages apprentices to acquire proficiency skills by self-training (Singleton, 1989). Here, a triple system design, which is one extension of the lens model, provides us with useful information for identifying the characteristics of this apprenticeship system (see Figure 4).

This method is called a *triple system design* because it analyzes cues from three directions for two judgments and the environment and is expected to explain that plural participants cooperatively judge by way of "interpersonal conflict" and "interpersonal learning" in the judgment processes between the two participants (see Figure 4). If one participant has a different ecological validity and cue utilization validity from another, the triple system design recognizes it as an interpersonal conflict and must solve this conflict through negotiations with mutual observations (this process of solving the conflicts is called *interpersonal learning*). Here, we need to change our own focus from the reciprocal relations between the participant and the environment to the reciprocal relationships between both participants, because the interrelations of each participant's utilization validities become more important in the case of interactions.

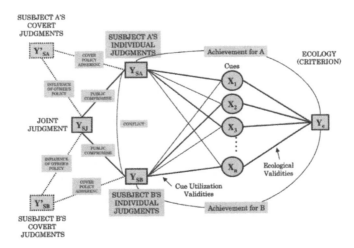

FIGURE 4 Triple system design extended to explain plural participant's judgment.

3.3. Skill-Transfer Depicted as the Triple System Design

In this section, we suggest that two difficulties of skill-transfer are expressed by two different interactions with the skill-transfer agent. The agent is requested to propose alternative tuning parameters to the user, and interactions between the agent and the user are depicted by the triple system design. The user (an expert in some cases and an apprentice in other cases) is requested to decide whether to accept the agent's proposals with regard to the operation. Because the agent updates its proposing rules through interaction with the user, the role of the agent is different depending on the degree of proficiency of each user.

At first, the agent is requested to interact with an expert who is proficient in tuning the parameters of image sensors (see Figure 5a). This process corresponds to one of extracting skills from the expert. This does not present a severe burden for the expert compared to the work of everyday tuning, because all the expert has to do is decide whether he or she accepts the agent's proposals.

Second, apprentices are assigned to the agent that has obtained proficient rules through interaction in the last phase (Figure 5b). What is important here is that it is assumed that the agent continues to update its own proposals even after the agent has already acquired tuning skills. As a result, the apprentice is expected to acquire his own individual utilization validities.

The detail mechanism of the interactive agent that achieves the previously mentioned two interactions is explained in the next section.

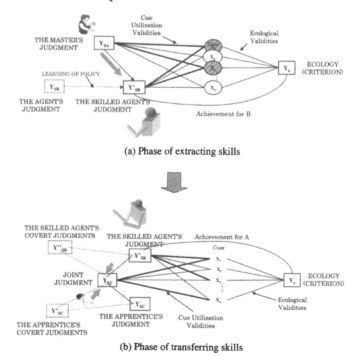

(a) Phase of extracting skills

(b) Phase of transferring skills

FIGURE 5 Different interactions mediated by triple system design.

4. DESIGN OF INTERACTIVE SKILL-TRANSFER SYSTEM

4.1. Appearance of the Interface System

Figure 6 shows the appearance of our interface. This interface has four features: an image viewer showing an image of the inspection objects; an operation viewer showing the agent's proposal; buttons for receiving the user's responses (i.e., acceptance or rejection of the agent's proposals); and a facial icon representing the agent's confidence status.

4.2. Design of the Rules for the Agent

In this article, we adopt an interactive learning classifier system (ILCS) as the agent's learning method for the first type of skills (i.e., the describable skills). The learning classifier system (LCS) is a well-known architecture proposed by Holland (Holland, Holyoak, Nisbett, & Thagard, 1986), and each classifier rule is evaluated by an evaluation function provided by the system designer in advance. Different from an ordinary LCS, each classifier in the ILCS is evaluated by users who are requested to interact with the system (e.g., Katagami & Yamada, 2002). Our interactive agent monitors an object image taken by the image sensor and proposes alternative operations to change the location and size of the inspection frame within the image using classifier rules that are learned by observing the user's performance. Figure 7 shows an example image of various operations to change the location and size of the frame. Figure 8 shows the system flow of interaction with the agent.

1. The agent shows an image of the inspection objects.
2. The agent segments the image in the inspection frame into nine image parts and divides the brightness of these parts into three different levels (dark, middle, and bright).

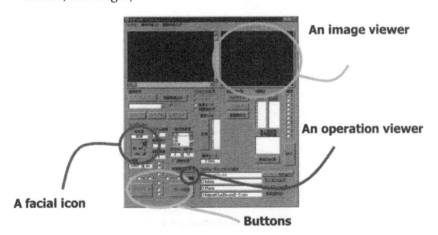

FIGURE 6 The interface system.

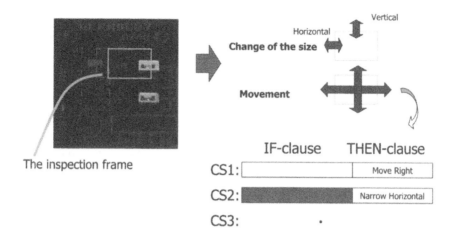

FIGURE 7 An example of operations.

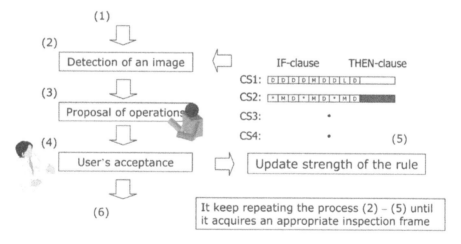

FIGURE 8 The system flow of ILCS.

3. The agent decides the provisional proposal of the tuning parameters and shows this proposed through the operation viewer. It is depicted as one of 12 different operations (to move up, down, right, and left and to widen or narrow its size in each direction).
4. If the user agrees with the agent's proposals, the agent increases the confidence in the related classifier rule by updating the numerical strength assigned to that rule.
5. The user and the agent repeat this cycle until the inspection frame satisfies the given end condition.

When the agent cannot find any classifier rules whose if-clause matches the situation, the agent is assumed to add a new classifier rule by abstracting the situation to the if-clause form and by combining it with new operations randomly (this process is a typical method of evolutional computation to create new rules, referred to as the Michigan approach by Holland, 1986).

4.3. Expression Using a Facial Icon

The agent is equipped with various icons of facial expression to present how confident it is in its proposals to human users. Such facial expressions allow the user to interact with the system more intuitively and interactively (Bisantz, Finger, & Llinas, 1999; Maes, 1994). Figure 9 shows the transition map from the past facial icon to the next facial icon. For instance, the agent reacts with the happy expression when the user accepts the agent's confident proposals. On the other hand, the agent reacts with the sad expression when the user rejects the agent's confident proposals. The effectiveness of using these facial icons is discussed in section 6.2.

It is important here that users need not always obey the agent's instructions. Rather, they can select alternative operations based on their own judgment or through attempts to see the responses of the agent to his or her temporal judgment. These intuitive reactions are expected not only to motivate the user to interact more with the agent him or herself, but also to enlighten the users themselves to explore their way of performing the tuning parameter tasks. We think this kind of continuous and reflective interaction effectively encourages users to demonstrate and actively self-discover the tacit aspects of human skills that cannot be captured by classifier rules.

4.4. Generalization of Classifier Rules

It is necessary to extend any learning rules acquired by the agent, because the quality of the tuning rules depends on both the quality and the quantity of the agent's experience in tuning various inferior goods with the user.

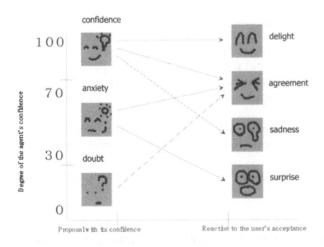

FIGURE 9 Transition mapping of facial icons.

It is well known that conditional parts of classifier rules have the special attribute #, known as "don't care," which means that the rule can fire in any situation with regard to the concerned attribute. As a result, attribute # is expected to extend the scope of application for the classifier rule. Usually, this attribute is added into the conditional parts of the classifier rules at random and is ineffective because it evaluates the validity of the rules only as a result of an evolutional computation mechanism. In this study, ID3 (Quinlan, 1979), which is a famous machine learning architectures, is introduced to add attribute # arbitrarily into the conditional parts of the classifier rules. ID3 gives our proposing algorithm effective generalization that can add attribute # selectively to its conditional parts of the classifier rules.

Figure 10 shows the pre-experiment to confirm the effectiveness of the generalization by the ID3 algorithm. The user is requested to tune parameters with our proposing agent for a certain inspection object (to pick up inferior connectors with broken pins). The x axis represents each step of the tuning process and the y axis represents the transition of the barycentric interval and boundary interval. Figure 10a shows the transition of both intervals by rote learning, which memorizes the conditional part as experienced by the agent, and Figure 10b shows the transition by generalization with ID3. Comparison of the two approaches demonstrates that generalization with ID3 provides criteria that distinguish inferior goods from all goods more stably.

5. THE FIRST EXPERIMENT: DIFFERENCES BETWEEN EXPERT AND APPRENTICE

5.1. Different Participants and Target Objects

In this study, the following two types of participants with different skill levels were used.

The expert: A skilled worker who has been engaged in maintenance work for 15 years and is well versed in image processing.

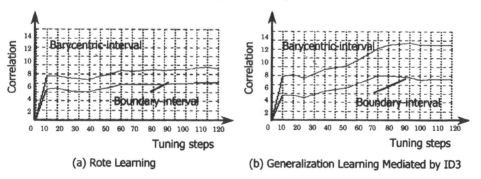

(a) Rote Learning   (b) Generalization Learning Mediated by ID3

FIGURE 10 Effectiveness of introducing ID3.

The apprentice: A bachelor student of our university at the time of the experiments. He is not familiar with an image processor, and only knows the objective of the image sensor.

The two participants were requested to regulate the parameters of the image sensor for distinguishing inferior goods from the following three objects: business cards, connectors, and Integrated Circuits (IC). Figure 11a and 11b shows example images of inferior goods for connectors with broken pins and ICs with broken pins. In all cases, it was difficult for the apprentice to regulate the image sensor sufficiently enough to appropriately detect inferior goods.

5.2. Tuning Parameters Mediated by the Interactive Agent

The expert and apprentices are first requested to regulate the parameters of the image sensors by interacting with the agent. The first target is assumed to be dirty business cards. Here, the user (sometimes the expert, sometimes the apprentice) is requested to continue selecting operations of tuning parameters until the system acquires the appropriate inspection frame. This set of tuning steps is called "a trial," and each user is requested to conduct 10 trials in each experiment.

Figure 12 shows an example transition of the agent's confidence in proposing alternative operations, and the transition of differences between the agent's proposal and the apprentice's choice. This transition is the 11th trial after the user (the user is the apprentice in Figure 12) interacts with the agent in 10 trials. Therefore, the agent's high confidence is because it has already experienced several interactions with the user. The x axis represents each step of the tuning process and the y axis represents the transition in each step.

Here, the differences between the agent's proposal and the user's selection are divided into four levels and are defined as follows. The value of zero indicates that the user accepts the agent's proposals as it is. On the other hand, the higher the number is, the more strongly the user rejects the agent's proposal.

What is interesting here is that the transition of the agent's confidence does not draw a simple learning curve, but rather a complex one, as shown in Figure 12. For instance, the A mark in Figure 12 indicates an interesting point where the transition of the agent's confidence fluctuates although the user agrees with the agent's pro-

(a) Connector with Broken Pins   (b) ICs with Broken Pins

FIGURE 11 Example images of inferior goods.

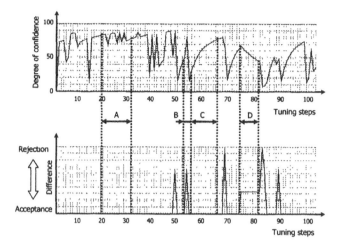

FIGURE 12 The transition of the agent's confidence and the difference between the agent's proposals and the user's choice.

posals. This fluctuation mandates that the agent needs to produce new conditional parts of rules for situations where the agent is less experienced.

On the other hand, the B mark indicates the situation where the agent's confidence drops off suddenly. This occurs when the user strictly rejects the agent's proposals or the agent must produce a new rule. The C mark signifies that such a new rule has gradually gained confidence through the user's acceptance. The D mark represents another situation where the confidence value decreases gradually. Such declination indicates that the user does not strongly reject the agent's proposals.

5.3. Generalized Skills

When the target inspection object is changed, the transition of interaction between the agent and the user is expected to become more complex. The participants are requested to regulate the parameters of the image sensor when the target inspection objects are changed. The target inspection objects are therefore changed from the first object (business cards with dirty marks) to the second one (connectors with broken pins), and from the second one to the third one (ICs with broken pins).

Figure 13a and 13b shows the transition of the differences between the agent and the expert or apprentice, respectively (this graph indicates the transition of the 11th interaction). Comparing Figure 13a with that of 13b demonstrates that the expert requires a smaller number of steps to specify an appropriate inspection frame than the apprentice does. Additionally, the number of times the expert rejects the agent's proposals is fewer than that of the apprentice although the target inspection objects are changed while tuning the parameters. This result suggests that the expert obtained more general rules of tuning the parameters of the image sensor than the apprentice did.

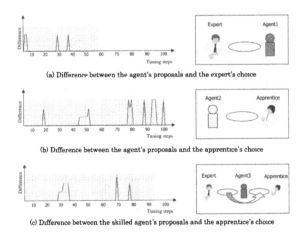

(a) Difference between the agent's proposals and the expert's choice

(b) Difference between the agent's proposals and the apprentice's choice

(c) Difference between the skilled agent's proposals and the apprentice's choice

FIGURE 13 Effectiveness of skill-transfer mediated by interactive agent.

5.4. Coherency of Judgment: Consistency

Discussions of the lens model present two types of indexes to validate the appropriateness of a judgment: "consistency" and "cognitive control." Consistency indicates how coherently the user can judge from the same sets of cues (Cooksey, 1996). On the other hand, cognitive control indicates how faithfully the user can judge against his or her own policy. Here, the "policy" is akin to the principles for the user to judge according to his or her experiences. These indexes distinguish the expert's works from the apprentice's.

Here, we can predict the agent's decision along with his or her consistency if we pay attention to the average confidence value of the total classifier rules, because the reinforcement value of the classifier rules remains low when the user cannot choose coherent operations from the agent's proposals. Figure 14 shows the transi-

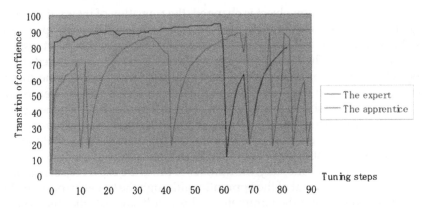

FIGURE 14 Transition of confidence (comparison of the expert with that of the apprentice).

tion of the agent's confidence in proposing operations during interaction with the expert and with the apprentice, respectively. Table 1 shows an average value of the agent's confidence in proposing operations through a series of interactions.

With Cooksey's (1996) discussions, these results hint that agent interaction with the expert achieves a higher consistency than that of the apprentice. Thus, introducing ILCS instead of the usual linear analysis model does not eliminate the difference between the expert and the apprentice.

5.5. Matching Property of Judgment: Cognitive control

The experiment in section 5.3 shows "cognitive control," which means how faithfully the expert and apprentice can judge against their own policy. Table 2 shows that the average confidence in proposing operations and the average number of acquired rules is arranged around the agent who formed it with the expert and the agent who formed it with the apprentice, respectively.

Because the user was requested to tune parameters for different inspection objects, it can be easily assumed that a new rule will be formed for each inspection object. Contrary to the usual conjecture, the agent who had interaction with the expert acquired only 26 rules. This is less than one fifth the number of rules acquired when the apprentice had interaction with the agent. If the sets of rules that the agent acquired by interacting with the user are general and independent from the superficial features of each inspection object, the number of rules should be small and the confidence level of proposing operations should be large. Comparison of the number of consequent rules provides possibilities for guessing how faithfully the expert judges along with his or her conviction, since consequent numbers of rules represent how comprehensively those rules were used in the evolutional computation.

Drawing from these discussions of consistency and cognitive control, the ILCS can be used to explain the features of proficiency skills that are addressed in the lens model studies. Although the ILCS was introduced instead of the linear analysis model because (a) environmental criteria cannot be obtained one at a time and (b) a phased change of the processes of acquiring skills cannot be expressed by the

Table 1: Difference Between the Expert and the Apprentice by Paying Attention to Consistency

	Interaction With the Expert	Interaction With the Apprentice
Average value of confidence	79.634	63.5

Table 2: Difference Between the Expert and the Apprentice by Paying Attention to Cognitive Control

	Interaction With the Expert	Interaction With the Apprentice
Average value of confidence	76.032	61.026
Average number of rules	26	136

usual linear analysis model, the ILCS can substitute for the lens model's expressiveness as regards proficiency skills.

6. THE SECOND EXPERIMENT: SKILL-TRANSFER

6.1. Results of Skill Transfer

In the second experiment, the expert tries to transfer his skills to the apprentice using our interactive agent system. At first, the expert interacts with the agent five times to inspect three types of work in turn. Then, the apprentice interacts five times with the agent to inspect the same set of three works in turn. Figure 13c shows the difference between the user's choice and the agent's proposal: (a) difference between the expert's choice and an inexperienced agent's proposals; (b) difference between the apprentice's choice and an inexperienced agent's proposals; (c) difference between the apprentice's choice and the experienced agent's proposals.

Figure 13a shows that the expert hardly rejects the agent's proposals through inspecting Work C, though the agent acquires classifier rules through inspecting other works (A and B). On the other hand, the apprentice could not help rejecting the agent's proposals during the inspection of Work C (see Figure 13b). These results mean that the agent that interacted with the expert could have acquired general rules independent from the difference in work, while the agent that interacted with the apprentice acquired only skills peculiar to each object. The fact that the agent that was educated by the apprentice has five times more rules than the agent educated by the expert supports this conclusion. Finally, Figure 13c illustrates the difference of the operations between the apprentice's choice and the experienced agent's proposal. As a result of this interaction, the frequency of the apprentice's rejection and the number of steps needed to finish the tuning decreases.

These results can be also verified by paying attention to the transition of the barycentric interval and the boundary-interval. Figure 15 shows the transition of the barycentric intervals for Figure 13b and 13c, respectively (the transition of the boundary interval is omitted because it is similar to the transition of the barycentric

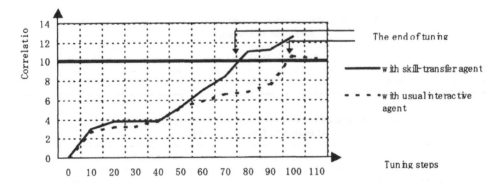

FIGURE 15 Improvements in learning rules by skill-transfer.

interval). Here, the end condition for completing the regulation of the parameters for the image sensor is set into 10 or more of the correlation values. Figure 15 clearly shows that the agent who had interaction with the expert finished regulating the parameters in a smaller number of steps than the agent who had interaction with the apprentice. The previously mentioned results demonstrate that our proposal system can improve apprentice performance mediated by an interactive agent.

6.2. Discussion

After the experiments, we asked the apprentice to fill out a questionnaire describing his impressions of interacting with the agent and of his own task performance. The following is a summary of the representative comments.

> I had expected interactions with the agent to be amazing and might have put excessive confidence in the agent's proposals, because I had known that the agent had been educated by the expert. However, I could not understand the experienced agent's proposals at the beginning of the session, because I did not know what parts the agent was paying attention to. Especially in the final steps of the session, I often favored choices different from the agent's, because I did not know what the sub-goal for the agent to finish inspecting was. That is, the goals sought by the agent were invisible to the human user.

The previous examinations illustrate that the typical difference between the expert and the apprentice is related to the method of changing the size of the inspection frame. The apprentice is apt to proceed with tuning by changing the size of the inspection frame, whereas the expert pays more attention to the movements of the inspection frame and a decision on the size of the frame is made at an early stage. This decision does not change during the session. Additionally, the apprentice's impression, "I was urged to pay attention to relations among operations by that system," supports this finding. The expert glances at the presented image less frequently than the apprentice does, which means that the expert pays more attention to the earlier steps of the operation.

7. CONCLUSIONS

In this article, we analyzed the difficulties of skill transfer from an ecological psychology viewpoint and designed an interactive skill transfer agent based on the apprenticeship system. Both the expert who engaged in image processing and the apprentice inspected objects in practice by interacting with the agent. According to the answers to our questionnaires, the interactions with the agent were so naturalistic that both the expert and the apprentice were not conscious of being engaged in a task of skill transfer. Answers to the questionnaires also showed that the apprentice obtained skills that are independent from the objects of the inspection. We can conclude that our system succeeded in solving the earlier mentioned two difficulties: low-cost extraction of skills from the expert and autonomy for apprentices to learn skills by themselves.

In this article, we introduced the viewpoint of ecological psychology that regards reciprocal relations between a participant and an environment as important. It should be noted that skills that should be transferred are defined not by only the instructor but by the reciprocal relationships between the instructor and the successor. In future work, we expect to specify how such reciprocal relationships are constructed through interaction between the instructor and the successor.

REFERENCES

Bisantz, A. M., Finger, R., & Llinas, J. (1999). Human performance and data fusion based decision aids. *Proceedings of the 2nd International Conference on Information Fusion-Fusion'99, International Society of Information Fusion.*

Boring, E. G., & Harper, R. S. (1948). Cues. *American Journal of Psycology, 101,* 112–123.

Brunswik, E. (1956). *Perception and the representative design of psychological experiments.* Berkeley: University of California Press.

Clancey, W. J. (1983). The advantages of abstract control knowledge in expert system design. *Proceedings of AAAI-83,* 74–78.

Cooksey, R. W. (1996). Judgment analysis—Theory, methods, and applications. San Diego, CA: Academic Press.

Davis, R. (1979). Interactive transfer of expertise: Acquisition of new inference rules. *Artificial Intelligence, 12,* 121–157.

Holland, J. H. (1986). Escaping brittleness: The possibility of general-purpose learning algorithms applied to rule-based systems. In R. S. Michalski, J. G. Carbonell, & T. M. Mitchell (Eds.), *Machine learning: An artificial intelligence approach* (Vol. 2, pp. 593–623). San Mateo, CA: Morgan Kaufmann.

Holland, J. H., Holyoak, K. J., Nisbett, R. E., & Thagard, P. R. (1986). *Induction.* Cambridge, MA: MIT Press.

Katagami, D., & Yamada, S. (2002). Interactive evolutionary robotics from different viewpoints of observation. *IEEE/RSJ International Conference on Intelligent Robots and Systems (IROS-2002),* 1108–1113.

Maes, P. (1994). Agents that reduce work and information overload. *Communications of the ACM, 37*(7), 30–40.

Norris, R. (1993). *The musician's survival manual: A guide to preventing injuries in instrumentalists.* International Conference of Symphony and Poera Musicians (ICSOM).

Quinlan, J. R. (1979). Discovering rules by induction from large collections of examples. In D. Michie (Ed.), *Expert systems in the micro electronic age.* Edinburgh, England: Edinburgh University Press.

Rogoff, B. (1990). *Apprenticeship in thinking: Cognitive development in social context.* Oxford, England: Oxford University Press.

Shortliffe, E. H. (1974). *MYCIN: A rule-based computer program for advising physicians regarding antimicrobial therapy selection.* Stanford, CT: Stanford University.

Shortliffe, E. H. (1976). *Computer-based medical consultations: MYCIN.* New York: Elsevier.

Singleton, J.(1989). *Japanese folkcraft pottery apprenticeship: Cultural patterns of an educational institution in coy ed.* Albany: State University of New York Press.

Ueno, K., & Furukawa, K. (1999). Postual effects during cell performance. *Proceedings of the 2nd International Conference on Cognitive Science,* 1132–1135.

INTERNATIONAL JOURNAL OF HUMAN–COMPUTER INTERACTION, *17*(1), 89–102

Visualization of Respiration in the Embodied Virtual Communication System and Its Evaluation

Tomio Watanabe
Faculty of Computer Science and System Engineering, Okayama Prefectural University
CREST of Japan Science and Technology Corporation (JST)

Masamichi Ogikubo
Yutaka Ishii
Graduate School of System Engineering, Okayama Prefectural University

A proposed embodied virtual communication system provides a virtual face-to-face communication environment in which two remote talkers can share embodied interaction by observing their interaction with two types of avatars. One is VirtualActor, a human avatar that represents talker communicative motion and respiratory motion. The other is VirtualWave, an abstract avatar that expresses human behavior and respiration by simplified wave motion. By using the system for the analysis by synthesis of embodied communication, the effectiveness of the visualization of respiration in VirtualActor and VirtualWave is demonstrated by the analysis of the entrainment of interaction and the sensory evaluation in remote communication.

1. INTRODUCTION

Humans communicate smoothly with nonverbal behavior such as nodding and gestures as well as verbal messages. In particular, the coherently related synchrony of independent biorhythms among verbal and nonverbal behavior between talkers, which is called *entrainment* in communication, plays an important role in human interaction and communication (Watanabe, 2001). The phenomenon is observed in mother–infant interaction as a primitive form of communication (Condon & Sander, 1974; Kobayashi, Ishii, & Watanabe, 1992). Entrainment in communication is also observed in physiological indexes such as respiration and heart rate variability (Watanabe, Okubo, & Kuroda, 1996). This embodied communica-

This work under our project E-COSMIC (Embodied Communication System for Mind Connection) has been supported by CREST of JST.

Requests for reprints should be sent to Tomio Watanabe, Faculty of Computer Science and System Engineering, Okayama Prefectural University, 111 Kuboki, Soja, Okayama 719–1197, JAPAN. E-mail: watanabe@cse.oka-pu.ac.jp

tion closely related to behavioral and physiological entrainment is an essential form of communication that forms the relation of interaction between talkers through mutual bodies (Ogawa & Watanabe, 2001; Watanabe, Okubo, & Ogawa, 2000). Hence, the clarification of this mechanism would be founded on the development of human-friendly communication systems.

The analysis and synthesis of nonverbal behavior and the development of communication systems using avatars have been carried out (Bullinger, Ziegler, & Bauer, 2002; Choi, 1996; Morishima, 1999; Nakatsu, 1998). However, to clarify embodied communication systematically, the development of a virtual face-to-face communication system is indispensable in which talkers' behavior is freely controlled in real time in a virtual communication environment.

We already proposed a concept of such a virtual face-to-face communication system for human interaction analysis by synthesis, and the prototype of the system was developed (Watanabe, Okubo, & Inadome, 1998). By using the system, two remote talkers can communicate observing their interaction of virtual actors representing themselves by sharing their embodiment in the same virtual space. On the basis of the system, an embodied virtual face-to-face communication system with two types of avatars for human interaction sharing was developed. One was VirtualActor (VA), a human avatar that represents human behavior more precisely than our previous system. The other was VirtualWave (VW), an abstract avatar in which the rhythm of human interaction was expressed as a wave generated by head motion to clarify an essential role of interaction (Watanabe & Okubo, 1999; Watanabe, Okubo, Ishii, & Nakabayashi, 2000).

On the other hand, physiological entrainment such as respiration as well as nonverbal entrainment is confirmed in face-to-face communication, which would be closely related with the sharing of embodiment in communication (Watanabe & Okubo, 1997). In particular, respiration as the physiological main factor would play an important role in assisting human interaction and communication because the burst-pause of speech is closely related with respiration, and the importance of respiration is pointed out in empirical human interaction such as counseling and the martial arts (Watanabe & Okubo, 1998). Therefore, the application of respiration to the system by visualizing respiration explicitly would contribute to the functional evaluation of respiration and the analysis of the effects on interaction. In the development of our previous system, however, the introduction to the system of physiological information such as respiration was only proposed in the concept.

In this article, an embodied virtual face-to-face communication system with the function of visualization of respiration is developed in which two types of avatars are refined with VA representing human communicative motion and respiratory motion, and VW expressing human behavior and respiration by simplified wave motion. This is based on the assumption that the explicit representation of respiration in VA and VW is useful for sensing interactive rhythms in human communication. By using the system for the analysis by synthesis of embodied communication, the effectiveness of the visualization of respiration in VA and VW is demonstrated by the analysis of the entrainment of interaction and the sensory evaluation in remote communication.

2. EMBODIED VIRTUAL FACE-TO-FACE COMMUNICATION SYSTEM

2.1. Concept

The concept of the embodied virtual face-to-face communication system is shown in Figure 1. In the figure, VA is an interactive avatar that represents the talker's interactive behavior such as gestures, nodding, blinking, facial color, and expressions, paralanguage, respiration, and so forth on the basis of its verbal and nonverbal information and physiological information in a virtual face-to-face communication environment. Two remote talkers can communicate through their VAs, including one's own VA and the other VA, and get interaction awareness from the embodied interaction of VAs in the same virtual communication environment from any viewpoint. The analysis by synthesis for interaction in communication is performed by processing the behavior of VAs such as cutting or delaying the motion and voice of VAs in various conditions of the spatial relations and positions of VAs. For example, to examine the effects of only nodding on interaction, it is possible for a VA to represent only nodding without body motion even if the talker nods with body motion. Thus, the characteristics and relations of embodied interaction between talkers are systematically clarified by the analysis by synthesis of interaction in communication using the system where talkers are the observers of interaction as well as the operators of interaction through their VAs. Physiological measurements, such as respiration, heart rate variability, and facial skin temperature, as indexes of emotional states in communication are also utilized not only for evaluating the interaction quantitatively, but also for transmitting talkers' emotional change by the VA affect display in which facial color and expressions are synthesized on the basis of the measurement (Kuroda & Watanabe, 1998). By using the proposed system with the function of visualization of respiration, talkers can get interactive rhythms more effectively than our previous system, which represents only communicative motions.

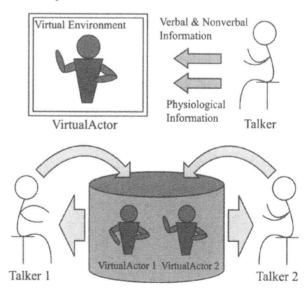

FIGURE 1 Concept of the embodied virtual communication system.

2.2. VA

An example of a virtual face-to-face scene with two VAs from the diagonal back-ward viewpoint of one own VA is shown in Figure 2, which was generated by using a virtual reality development software (SENCE8 WorldToolKit) and a Windows NT workstation (INTERGRAPH TDZ 2000 GT1) with a 3D graphic engine (Intense3D Wildcat 4000). The motions of the head, arms, and body for each VA are represented on the basis of the positions and angles measured by four magnetic sensors (Polhemus FASTRAK) put on the top of the talker's head, both wrists, and back. The talker's expiration and inspiration are measured by a thermal sensor in a Multi Telemeter System (NIHON KOHDEN, [WEB-5000], TR-511G) put on the entrance of the talker's nose. The VA's motion of respiration in correspondence to the measurement is also represented in real time by expansion and contraction of the chest and up-and-down motions of the shoulders, as shown in Figure 3. The frame rate to represent VAs is 24 f/sec, where the data of VAs motion and voice are transmitted through 100 Mbps Ethernet at the same time they are stored at the sampling rate of 24 Hz and 11,025 Hz, respectively, in each computer system.

FIGURE 2 Example of virtual face-to-face scene with two VirtualActors (VAs) that represent his or her own self and the partner.

Expiration   Inspiration

FIGURE 3 Example of VA motion of respiration.

2.3. VW

Although we have made a human VA representing human behavior precisely, we have also made an abstract avatar called VW in which the communication function of the VA is simplified as a wave motion constructed by 6 × 6 cubes to clarify an essential role of interaction. The VW rhythm to characterize interactive rhythm and behavior is represented by only the head motion measured by one magnetic sensor put on the top of the talker's head. This is because the head motion includes the essential action of regulator for the flow of conversation such as nodding by which each talker discriminates his or her VW from the partner VW and shares their interaction, as the effects of head motion on the VA interaction has already demonstrated (Ishii & Watanabe, 2002). Nodding is expressed by an up-and-down quake of cubes in which the wave approaches a quadrangular pyramid in shape. The VW motion of respiration is expressed by concentration and diffusion of cubes corresponding to expiration and inspiration respectively, measured by the same method for VA, as shown in Figure 4. The horizontal shift such as back and forth is expressed by the parallel displacement of cubes in proportion to the shift. The frame rate to represent VW is 30 f/sec in the same environment as VA.

3. COMMUNICATION EXPERIMENT

3.1. Experiment by Using VA

Experimental method. The setup of the experiment in which two remote talkers converse in separate rooms is shown in Figure 5. The situation of the communication experiment such as the correspondence of motions between talkers and VAs was recorded through a video editor (SONY FXE-100) by two video cameras, as shown in Figure 6. The angles and positions measured by four magnetic

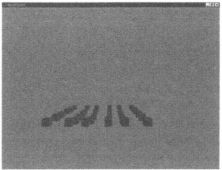

Expiration   Inspiration

FIGURE 4 Example of VirtualWave (VW) motion of respiration.

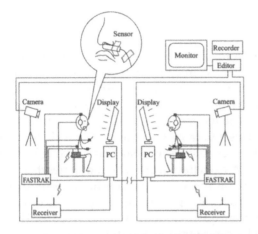

FIGURE 5 Experimental setup.

FIGURE 6 Example of communication scene.

sensors and the respiration data through an A/D converter by the thermal sensor were stored at the sampling rate of 24 Hz. At the same time, the motion and respiration data of the VAs and the voice data of the talkers were directly transmitted through 100 M bps Ethernet. The experiment was performed under two communication scenes of both VAs including one own and the partner (2VAs) and only the partner VA (1VA), as shown in Figure 7. The experiment consisted of three cases under the conditions of the scene and the visualization of respiration. Case 1 was examined every three minutes under two scenes without the visualization of respiration. Case 2 and case 3 were examined under the conditions whether the respiratory motion of the VA was visualized or not under the scene of 1VA and 2VAs respectively. The experimental time was set at 3 min for the respiratory motion, and 2 min without respiration, because the VA without respiratory motion had been experienced in case one. The experiment was performed for about 20 min including the questionnaire per pair in eight pairs of 16 Japanese students who were familiar each other. After the experiment in each case, sensory evaluation by the questionnaire about the system was examined with 7-point bipolar rating scales from –3 (*not at all*) to 3 (*extremely*) in which the score 0 denotes *moderation*, from the viewpoint of sharing the embodiment for communication support. In the

FIGURE 7 VA communication scenes in the experiment. 2VAs: Diagonal view with two VAs including his or her ownself and the partner. 1VA: Frontal view with the partner VA to talk to.

remote communication experiment, talkers made conversation about their everyday lives, hobbies, and so forth.

Result of sensory evaluation.   Figure 8 shows the result of sensory evaluation: the evaluation of 2VAs based on 1VA in Case 1, the evaluation of 1VA with respiration based on 1VA without respiration in Case 2, and the evaluation of

> i: Enjoyment   ii: Interaction   iii: Usage   iv: Liveliness
>
> v: Preference   vi: Sharing/Separation   vii: Unification
>
> viii: Relief/Uneasiness   ix: Relaxation/Tension
>
> x: Naturalness   xi:Presence   xii: Familiarity   xiii: Organic

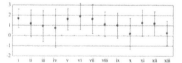

Case 1: Evaluation of 2VAs based on 1VA.

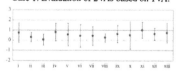

FIGURE 8 Sensory evaluation for VA: Case 1: Evaluation of 2VAs based on 1VA; Case 2: Evaluation of 1VA with respiration based on 1VA without respiration; Case 3: Evaluation of 2VAs with respiration based on 2VAs without respiration.

Case 2: Evaluation of 1VA with respiration based on 1VA without respiration.

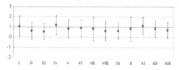

Case 3: Evaluation of 2VAs with respiration based on 2VAs without respiration.

2VAs with respiration based on 2VAs without respiration in Case 3. In any item of Case 1, the scene of 2VAs is more affirmatively evaluated than that of 1VA. This demonstrates the effectiveness of representation of the talker's own VA and the partner VA in the same virtual space, as proposed by our previous system. In particular, the effectiveness of sharing the embodiment through mutual VAs is confirmed by the highly affirmative evaluation in the items of sharing and separation and unification. In Case 2 and Case 3, because the evaluation of the VA with respiration in comparison with the VA without respiration is affirmative, the effectiveness of visualization of respiration is confirmed. In particular, the evaluation in Case 3 is more remarkably affirmative than that in Case 2. This demonstrates the effectiveness of the representation of the talker's own VA as well as the partner VA in the virtual space. In addition, the effectiveness of visualization of respiration is confirmed by the preferable evaluation in the items of enjoyment and preference.

Result of behavioral analysis. The interaction between talkers was analyzed on the basis of the relation between a talker's head motion and the other one in the communication experiment. On the basis of the angle p(i) measured at the sampling rate of 24 Hz by the magnetic sensor put on a talker's head, head motion x(i) was calculated from the difference between the following and previous data [p(i+1)-p(i-1)] and then defined as binary from the average threshold. The angle of head motion was restricted for the rotation of nodding, because nodding plays an important role in human interaction. Binarization eliminated noise and large data unrelated to essential interaction. The relation between a talker's head motion and the other one was evaluated by using the following cross-correlation function $C(\tau)$ in four scenes: 1VA without respiration, 1VA with respiration, 2VAs without respiration, and 2VAs with respiration.

$$C(\tau) = \frac{\sum_{i=1}^{n-\tau} \{x(i) - \mu_x\}\{y(i+\tau) - \mu_y\}}{\sqrt{\sum_{i=1}^{n} \{x(i) - \mu_x\}^2} \sqrt{\sum_{i=1}^{n} \{y(i) - \mu_y\}^2}}$$

Where m_x and m_y are the mean values of x and y respectively, n is the number of data, and τ is time delay.

Figure 9 shows an example of time changes of cross-correlations between a talker's head motion and the other one. The time changes of $C(\tau)$ within 2-sec time lag in 30-sec analyzing periods were expressed by the shading image whose analyzing periods were shifted by 1sec. The darkness denotes the intensity of positive correlation. From the typical cross-correlograms in the figure, strong positive peaks are noted at time lag 0. The positive correlation between the talker's head motions, which indicates the entrainment in human interaction, can be regarded as one of the indexes for the evaluation of effective communication.

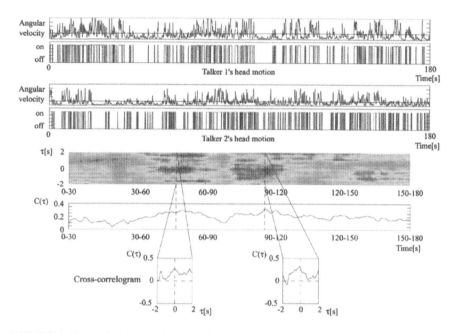

FIGURE 9 Example of time changes of cross-correlations between a talker's head motion and the other talker's head motion.

Figure 10 shows the mean and standard deviation of the positive peak of cross-correlations between the talkers' head motions in each scene from 16 participants. The effectiveness of visualization of respiration for the entrainment of interaction was demonstrated from the statistical result that there was significance with a significance level of 1% between 1VA with respiration and 1VA without respiration and between 2VAs with respiration and 2VAs without respiration. The effectiveness was also confirmed from the result of sensory evaluation in the previous section.

3.2. Experiment by using VW

Experimental method.   The communication experiment using VW was also examined on the basis of the talker's respiration measured by a thermal sensor put on the entrance of the talker's nose, and the talker's head motion measured by a magnetic sensor put on the talker's head in the same way as with the VA as shown in Figure 5. The frame rate to represent VWs was 30 f/sec when the data of the VWs motion and voice were transmitted through 100 Mbps Ethernet at the same time they were stored at the sampling rate of 30 Hz and 11,025 Hz, respectively. The experiment was performed under two communication scenes of both VWs including one own and the partner (2VWs) and only the partner VW (1VW) as shown in Figure 11. The experiment consisted of three cases under the conditions of the scene and the visualization of respiration. Case 1 was examined every 3 min under two scenes without the respi-

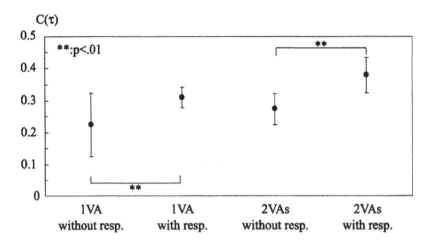

FIGURE 10 Mean and standard deviation of the positive peak of cross-correlations between talkers' head motions for VA.

ratory motion. Case 2 and Case 3 were examined under the conditions whether the respiratory motion of the VW was visualized or not under the scenes of 1VW and 2VWs. The experimental time was set at 3 min for the respiratory motion, and 2 min without respiration, in the same way as in the VA experiment. The experiment time was about 20 min including the questionnaire. The participants were the same eight pairs of 16 students as in the VA experiment. After the experiment in each case, sensory evaluation by the questionnaire about the system was examined in the same way as in the VA experiment. Everyday conversation was made in the experiment. The meaning of the VW motion was not explained to talkers.

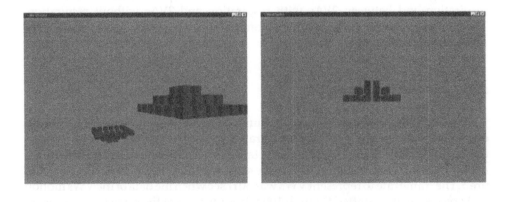

2VWs   1VW

FIGURE 11 VW communication scenes in the experiment. 2VWs: Diagonal view with two VWs including his own self and the partner. 1VW: Frontal view with the partner VW to talk to.

Result of sensory evaluation.   Figure 12 shows the result of sensory evaluation: the evaluation of 2VWs based on 1VW in Case 1, the evaluation of 1VW with
respiration based on 1VW without respiration in Case 2, and the evaluation of
2VWs with respiration based on 2VWs without respiration in Case 3. In Case 1, the
scene of 2VWs is more affirmatively evaluated than the scene of 1VW in the items
of enjoyment, preference, and unification. This demonstrates the effectiveness of
representation of the talker's own VW and the partner VW in the same virtual
space in the same result as the sensory evaluation of VA, as proposed by our previous system using VWs.

In Case 2 and Case 3, the evaluation of the VW with respiration in comparison
with the VW without respiration is affirmative in the items of liveliness and presence. The effectiveness of visualization of respiration is confirmed because talkers
can better see each other's embodied rhythms by representation of respiration than
in our previous system represented by only the talker's communicative motions.
The evaluation in Case 3 is more affirmative than that in Case 2, just as in the evaluation of the VAs. In any case, the effectiveness of visualization of respiration is demonstrated from the affective evaluation in the item of enjoyment. In the item of naturalness, however, the VW with respiration was given a low evaluation.

In the questionnaire, opinions for the communication effects of respiratory visualization such as reality for the presence of the partner and easiness for sensing the

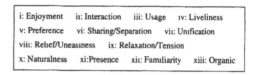

i: Enjoyment   ii: Interaction   iii: Usage   iv: Liveliness

v: Preference   vi: Sharing/Separation   vii: Unification

viii: Relief/Uneasiness   ix: Relaxation/Tension

x: Naturalness   xi:Presence   xii: Familiarity   xiii: Organic

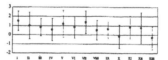

Case 1: Evaluation of 2VWs based on 1VW.

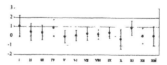

Case 2: Evaluation of 1VW with respiration based on 1VW without respiration.

FIGURE 12 Sensory evaluation for VW: Case 1: Evaluation of 2VWs based on 1VW;
Case 2: Evaluation of 1VW
with respiration based on
1VW without respiration;
Case 3: Evaluation of 2VWs
with respiration based on
2VWs without respiration.

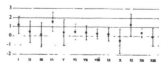

Case 3: Evaluation of 2VWs with respiration based on 2VWs without respiration

rhythms of interaction were described. This demonstrates the effectiveness of respiratory visualization for supporting human interaction.

Result of behavioral analysis. The interaction between talkers was analyzed on the basis of the talker's head motion in the communication experiment. Head motion data x(i) was calculated from the angles measured at the sampling rate of 30 Hz by the magnetic sensor put on a talker's head in the same way as in the VA experiment. The relation between a talker's head motion and the other one was evaluated by using cross-correlation function $C(\tau)$ in four scenes: 1VW without respiration, 1VW with respiration, 2VWs without respiration, and 2VWs with respiration. Figure 13 shows the mean and standard deviation of the positive peaks of cross-correlations between talkers' head motions in each scene from 16 participants. The positive correlation in 2VWs with respiration was higher than that in 2VWs without respiration from the statistical result that there was significance with a significance level of 1%. This demonstrates the effectiveness of the representation of respiratory motion for the entrainment of embodied rhythms. In addition, the effectiveness of the entrainment of the embodied rhythms by representation of both VWs for human interaction was confirmed from the statistical result that there was significance with a significance level of 1% between 1VW with respiration and 2VWs with respiration.

However, there was no significant difference between 1VA with respiration and 1VA without one. This indicates that there was no effect for the entrainment between talkers by whether the respiratory motion was visualized or not in the scene of 1VW. It was assumed that talkers could easily recognize the relation between the VW respiratory motion and the talker's respiration based on the representation of the talker's own VW in the scene of 2VWs, while they were hard to relate the VW respiratory motion to their own respiration in the scene of 1VW. In the previous section, because the sensory evaluation in Case 2 was affirmative in the items of

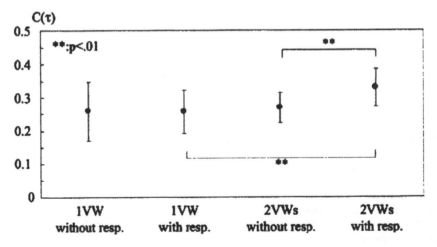

FIGURE 13 Mean and standard deviation of the positive peak of cross-correlations between talkers' head motions for VW.

liveliness and presence, the VW motion of respiration in 1VW was also effective in presenting embodied rhythm.

4. CONCLUSIONS

In this article, an embodied virtual face-to-face communication system with the function of visualization of respiration was developed by using two types of avatars: one was VA, a human avatar that represents talkers' communicative motion and respiratory motion; the other was VW, an abstract avatar that expresses human behavior and respiration by a simplified wave motion. By using the system for the analysis by synthesis of embodied communication, the analysis of the entrainment of interaction and the sensory evaluation demonstrated the effectiveness of the visualization of respiration in VA and VW. In particular, the effectiveness of the entrainment of embodied rhythms and the feelings of liveliness and presence were confirmed.

Thus, this embodied communication system is quite unlike the conventional telecommunication systems in which the communication space between talkers is separated because only the partner who is talking is displayed. The system would form the foundation of information media and communication technologies as well as the methodology for the analysis and understanding of various human interactions and communications.

REFERENCES

Bullinger, H.-J., Ziegler, J., & Bauer, W. (2002). Intuitive human–computer interaction—Toward a user-friendly information society. *International Journal of Human–Computer Interaction, 14*, 1–23.

Choi, S. (1996). Analysis of communication behaviors in ISDN-TV model conferences using the synchronous and asynchronous speech transmission. *IEICE Transactions on Communications, E79-B*(6), 728–736.

Condon, W. S., & Sander, L. W. (1974). Neonate movement is synchronized with adult speech: Interactional participation and language acquisition. *Science, 183*, 99–101.

Ishii, Y., & Watanabe, T. (2002). Analysis by synthesis of embodied communication in which the head motion of listener VirtualActor is inconsistently stopped. *The Transactions of Human Interface Society, 4*(3), 9–16.

Kobayashi, N., Ishii, T., & Watanabe, T. (1992). Quantitative evaluation of infant behavior and mother–infant interaction. *Early Development and Parenting, 1*, 23–31.

Kuroda, T., & Watanabe, T. (1998). Analysis and synthesis of facial color using color image processing. *JSME International Journal, 41*(2C), 307–312.

Morishima, S. (1999). Multiple points face-to-face communication in cyberspace using multi-modal agent. *Human—Computer Interaction, 2*, 177–181.

Nakatsu, R. (1998). Nonverbal information recognition and its application to communications. *Proceedings of 3rd International Conference on Face and Gesture Recognition*, 2–7.

Ogawa, H., & Watanabe, T. (2001). InterRobot: Speech driven embodied interaction robot. *Advanced Robotics, 15*(3), 371–377.

Watanabe, T. (2001). E-COSMIC: Embodied communication system for mind connection. *Usability Evaluation and Interface Design, 1*, 253–257.

Watanabe, T., & Okubo, M. (1997). Evaluation of the entrainment between a speaker burst-pause of speech and respiration and a listener respiration in face-to-face communication. *Proceedings of 6th IEEE International Workshop on Robot and Human Communication (RO-MAN 7)*, 392–397.

Watanabe, T., & Okubo, M. (1998). Physiological analysis of entrainment in communication. *Transactions of Information Processing Society of Japan, 39*(5), 1225–1231.

Watanabe, T., & Okubo, M. (1999). An embodied virtual communication system for human interaction sharing. *Proceedings of 1999 IEEE International Conference on Systems, Man, and Cybernetics*, 1060–1065.

Watanabe, T., Okubo, M., & Inadome, M. (1998). Virtual communication system for human interaction analysis. *Proceedings of 7th IEEE International Workshop on Robot and Human Communication (RO-MAN 8)*, 21–26.

Watanabe, T., Okubo, M., Ishii, Y., & Nakabayashi, K. (2000). An embodied virtual communication system with human Virtual Actor and abstract Virtual Wave. *The Transactions of Human Interface Society, 2*(2), 1–10.

Watanabe, T., Okubo, M., & Kuroda, T. (1996). Analysis of entrainment in face-to-face interaction using heart rate variability. *Proceedings of 5th IEEE International Workshop on Robot and Human Communication (RO-MAN 6)*, 141–145.

Watanabe, T., Okubo, M., & Ogawa, H. (2000). An embodied interaction robots system based on speech. *Journal of Robotics and Mechatronics, 12*(2), 126–134.

INTERNATIONAL JOURNAL OF HUMAN–COMPUTER INTERACTION, *17*(1), 103–124

Trends in Usability Research and Activities in Japan

Masaaki Kurosu
Research and Development Division, National Institute of Multimedia Education

Tadashi Kobayashi
Quality Assurance Department, Fujitsu Info Software Technologies

Ryoji Yoshitake
User Experience Design Center, IBM Japan

Hideaki Takahashi
Research and Development Division, National Institute of Multimedia Education

Haruhiko Urokohara
Daisuke Sato
Usability Research and Development Department, U'eyes Novas Inc.

This article presents an outline of information on the usability activities in Japan with a focus on the last 10 years. Although there were many activities in academia, substantial efforts were made in industry, and both activities coincided to form unique usability engineering in the country. Because of the language barrier that exists in many Japanese usability engineering professionals, just a few works have been presented at international conferences. This is the reason why the tried to summarize in English the usability engineering activity in Japan.

1. INTRODUCTION

This article describes research into usability and related practical activities that have been undertaken in Japan over the course of the past 10 years or so.

Here in Japan, usability research and related activities have only really taken off in recent years with the influence of ISO13407, although there had been steady activity in the fields of ergonomics and cognitive engineering even before that. Re-

Requests for reprints should be sent to Masaaki Kurosu, Research and Development Division, National Institute of Multimedia Education, 2–12, Wakaba, Mihama-ku, Chiba-shi, Chiba 261–0014 Japan. E-mail: PFD00343@nifty.com

cently it has become normal for companies to establish a dedicated usability section and install a usability laboratory. Although usability activities do not yet stretch to cover all products, it is safe to assume that such activities will only increase with time.

Domestically, these research and practical activities are the result of work by groups such as the Human Interface Society and the Japan Ergonomics Society. Unfortunately, due to language difficulties, the activities described herein are only partially known overseas.

2. TRENDS IN USABILITY RESEARCH

2.1 Research Trends in Usability Processes

In Japan, research papers on the incorporation of the results of usability testing into finished products began to appear only a few years ago. Although some manufacturers reportedly included usability activities in their product development procedures, almost no academic activity in this field is known to us before 1990. Since then, however, Amemiya (1990) and Ohtsuka (1992) reported on the application of principles and guidelines for usability and the user interface to general product design. Additionally, Y. Shibata (1992) proposed a user participation method and Tsuchiya, Horii, Kitamura, and Tamura (1996) proposed a design method whereby usability was the goal of product design solution.

Subsequently, as usability became a more readily recognized concept, human-centered design (HCD) and user-centered design (UCD) processes came to prominence and, around the time that ISO 13407 took effect in 1999, the concept of usability was framed in the HCD process (Arai, 2001; Igeta, 2001). Kurosu et al. (2001) took the lead in promulgating ISO13407 by writing an ISO 13407 book for the first time in the world, and recently (Bevan, Earthy, & Kurosu, 2001; Kurosu, 1997; Kurosu et al., 2001) research papers have appeared based on activities aimed at establishing HCD and UCD. Kurosu et al. (1999) also published a book promoting a new concept of user engineering.

Also during the 1990s, related works by Norman (1988), Landauer (1995), Nielsen (1993), and Cooper (1999) were translated into Japanese and published. Subsequently, Norman's entire series of writings on the subject was also translated. Japanese writers remained active in this field with notably Kurosu et al. (twice in 2001) and Kurosu (as editor in 2003) publishing works on usability, while Yamaoka, Suzuki, and Fujiwara (2000) promoted the interface design method known as SIDE, and Mitsubishi Electric Corporation (2001) reported on in-company activities.

Research in this field includes the work of Nagamatsu, Yamamoto, Iwata, and Yoshikawa (2001), who, based on the method of user-centered design and evaluation of virtual environments proposed by Gabbard et al., created a cooperative machine maintenance training system at a nuclear power plant and investigated the effectiveness of user task analysis, design and evaluation, and user-centered evaluation. Yoshitake and Hosokawa (2001) argued the case for UCD activities with their report on the application of user-centered design to the development of a consumer image processing software, which proved the effectiveness of task analysis, design walkthrough, and design evaluation and validation. Sato, Itoh, and Kobayashi (2002) succeeded in incorporating usability testing into the software development process and described their results. Other related publications include Hasegawa and Kobayashi (2001) on effective HCD and UCD activities; I. Okamoto, Tanaka, and Itoh (2001) on ongoing cases of continuous UCD activity; and Kanayama, Ogasawara, and Kimijima (2002) on a proposal for a new design framework by corporate Web site investigations. Additionally, Waida (2002), Arai , Itoh, Urokohara, and Takamoto (2002), and Wakamatsu, Sawada, and Nomoto (2002) reported corporate-level activities in HCD and usability.

In addition to the practical activities mentioned previously, Kajii and Itoh (2001) described the difference between usability testing for product evaluation and usability testing based on HCD, and Kurosu et al. (2001), Kurosu (2001), and Hosono, Inoue, Tomita, and Yamamoto (2002) published their results on research limited not only to Japanese users. Recently Nakakoji, Kamiya, Takashima, and Yamamoto (2001) proposed an approach and method for building interactive systems that are not only usable and useful, but also highly desirable, and Kurosu, Ito, Horibe, Hirasawa, and Fujiwara (2000) proposed COEDA (formerly SDOS), an interactive analysis method for analyzing approaches to HCD in businesses.

2.2 Requirements Method

Research into the gathering of information on user requirements, its analysis, and clarification of the conditions that determine those requirements has only become readily available in recent years.

Regarding the gathering of information on user requirements, S. Shibata (2000) set up a users' forum for the purpose of perfecting hitherto used methods of usability research such as usability testing and focus groups, site visits, Web surveys, and beta monitoring. This forum acts as a link between users and developers via media such as electronic bulletin boards and chat sites, mail magazines, mailing lists, and actual meetings, giving the currently registered 500 users an opportunity for their voice to be heard during the development process. With opinions such as, "This product is more expensive than I expected," and, "We should be careful that this does not cause consumers to lose trust in the company," expressed, this has proved to be an extremely beneficial source of information. Although not necessarily a systematic approach, many companies build and maintain a database of survey responses to actively gather information from users not just within their company, but also from external users.

By collecting user responses to questionnaires and usability testing, information on requirements early in the development process can be collected. For example, Ogino, Ichikawa, and Shimamura (2000) found benefits for creating product concepts by conducting questionnaire surveys and usability evaluation before the product concept was created. With the questionnaire survey, it was impossible to evaluate the concept and functionality of the system when it is in operation by users. For this type of testing, it was discovered that usability testing is more effective than the questionnaire survey. In addition, Mine, Ikeda, Inagaki, and Okuizumi (2001) described having assumed users use a product over a period of time and analyzed sequential changes in observations, levels of task accomplishment, and responses to questionnaires and interviews in order to analyze user requirements. The study confirmed positive benefits for later versions of the product by looking at the sequential change in the results, in terms of impression, level of mastery, usage, and needs. Mayuzumi et al. (2001) offered an example of a questionnaire-based approach. To attain effective communication on environment in a company's ecological site, they created a checklist based on usability evaluation items for the Web, the results of which were used to suggest a method of solution. They are now investigating how to extend these methods to communication design methods specialized for ecological sites.

Turning to the analysis of user requirements, model-based methods are now the most prevalent. Yamaoka and Baber (2000) developed a three-point task analysis system in which user information processing is judged and human error is estimated for each task based on three criteria: information gathering, understanding and decision-making, and operation. Yamaoka and Baber recommended combining this "micro" three-point task analysis method with a "macro" error estimation method known as Task Analysis for Error Identification (TAFEI), based on layered task analysis, the spatial scheme of state, and the transition matrix to give a comprehensive estimation of human error when a task is performed. The flow chart-based analytical method proposed by Urokohara (1996) illustrated a similar approach. Urokohara recommended the use of operational flow charts to regulate tasks and specifications in the development of interactive software. In this article, three types of operational flows were considered: thought flow, task flow, and screen transition diagram. This method was based on the thought flow as a representation of the user's dynamic thought patterns. Then the task flow is created to consider corresponding functions based on the thought flow. Finally screen transition was determined based on the task flow, following which screen design was carried out. In addition, Nakagawa and Mitome (2000) applied the evaluation grid interview method known as laddering (first developed by Sanui in 1986) and its analysis method to usability issues. In other words, starting from an ambiguous image of usability, they sought to find out to whom usability is of value and what that value is by expressing this information diagrammatically to clarify user benefits.

The model-based method is also applied in interface design for large-scale systems. Regarding such interface design, M. Takahashi, Ito, Niwa, and Kitamura (1999) argued the necessity of concrete design guidelines in order to precisely reflect the necessity of being able to model various task accomplishment modes and the importance of formulating and clearly communicating the requirements of system users. Based on this way of thinking, Nishikawa et al. (1999) proposed a method of functional adaptation for an interface comprised of two elements of user cognitive model in task accomplishment mode and clarification of user requirement specifications. The cognitive modeling would be based on an expert level of cognizance, from which the interface design requirements would be determined. Regarding the clarification of user requirements, matters not taken into account during the initial design phase would be determined and the interface would be gradually improved. The interface of the accelerator system used in experiments at the Fast Neutron Laboratory of Touhoku University was designed using this method.

For this kind of requirement analysis method, the contextual design method promoted by Bayer and Holtzblatt (1988) is attracting attention. Ito, Yamada, and Kurosu (2000) analyzed the effectiveness of the flow model construction method described as part of the contextual design method promoted by Bayer and Holtzblatt (1998) on qualitative information obtained during fieldwork and compared instances where ideas were generated after qualitative information was given to instances where ideas were generated after no qualitative information was given. Their results showed that giving or withholding qualitative information had no effect on the amount of ideas produced and that, content-wise, where possibilities were high, ideas on improvements were more likely to be forthcoming. Furthermore, Kurosu et al. (2001), regarding the process model required by ISO13407, recommended using an integrated method combining microethnography, contextual design, and scenario-based design for analyzing early processes in development. However, Sugizaki, Araki, Itoh, and Kurosu (2001) asserted that the flow model method used in contextual design was unreliable for major decisions due to differences in the way individuals draw flows, that chronological sequences were difficult to distinguish, and that it could not be applied to complicated systems. Moreover, Araki, Sugizaki, Itoh, and Kurosu (2001) identified several problems, such as the possibility of numerous variations in the sequential model method in contextual design, the conditional branches, displaying relations to other sequences, and describing sporadic actions, in highlighting the need for improvement in this method.

In addition, through Carroll's (1995; Rosson & Carroll, 2002) efforts, the scenario-based design method is also the subject of attention, and Itoh, Ikeya, and Nakano (2001) asserted that we should differentiate between "current scenarios" and "solution scenarios," where the solution scenario is based on problem analysis concurrent with the creation of the current scenario. Further to this, Oonishi and Go (2002) published in Japan details of a series of scenario methods.

2.3 Research Trends in Usability Evaluation

There is a high rate of usability evaluation activities among the usability activities under way in Japan. A broad spectrum of research activity is currently being carried out, including the study of existing evaluation methods, proposals for new evaluation methods, and the development of logging tools for actual operation. As many authors have commented, the persuasiveness of evaluation results over others is important in Japan, and a great deal of effort is being spent to improve the logicality of evaluation results and to quantitatively evaluate the resulting data. Although not included in this article, other evaluation activities than general-purpose researches are reported on various products such as portable information terminals, PC software, data broadcasting content, Web sites, equipment systems, and power shovels.

Regarding inspection evaluation methods, Kurosu and various research part-
ners proposed structured Heuristic Evaluation Method (sHEM) to strengthen the
structured level and efficiency of the problem finding abilities of existing HEM
(Kurosu, Matsuura, & Sugizaki, 1997; Kurosu, Sugizaki, & Matsuura, 1997, 1998;
Kurosu et al, 1999). This is useful as a guideline for usability novices to learn
heuristics. On the concept of cognitive walkthrough, Suzuki, Kaneko, and Ohmura
(1995) proposed Quick Usability Inspection System (QUIS), which improved the
speed of evaluation; Matsuura, Kurosu, and Sugisaki (1998) undertook a comparative study of usability testing and sHEM; Nakajima, Ichikawa, and Kato (2002) reported on research into supporting actual operations, and Kitajima, Blackman,
Polson, and Lewis (2002) proposed Cognitive Walkthrough for the Web (CWW)
and "Automated Cognitive Walkthrough for the Web."

Reports on usability testing evaluation methods include the proposal of a usability evaluation method allowing more active interaction with testers by physically
hiding a portion of the interface for a better understanding of the formulation process
of operational images (Takeuchi, 1997; Takeuchi & Miki, 1996), a proposal for an efficient observation logging method utilizing check boxes (Furuta & Sato, 2001), a report on the development of a remote evaluation system (Omata, Tanoue, Go, &
Imamiya, 2000), a report on usability evaluation testing on skilled users with significant amount of load (Harada & Shikano, 2002), and a report on the effectiveness of
testing with a pair of testers (Kawasaki, Kashimura, & Harada, 2000). Other proposals include an original Japanese questionnaire to handle subjective user evaluation
based on SUMI or SUS (T. Okamoto & Kurosu, 2000), a questionnaire method for Web
usability (Nakagawa, Suda, Zempo, & Matsumoto, 2001), and methods such as logging tools to make usability testing more efficient. In addition, Kurosu (1997, 1998)
proposed a Dual Task Model evaluation method for estimating operation times
when evaluating dual tasks, such as operating a car navigation system while driving.

Many methods have been proposed to improve the persuasiveness and ade-
quacy of usability evaluation results. Urokohara and colleagues proposed Novice
Expert ratio Method (NEM) to clarify problems by measuring the operation time
ratio of experts and novices (Kurosu, Urokohara, Sato, Nishimura, & Yamada,
2002; Urokohara, Furuta, Tanaka, & Kurosu, 1999; Urokohara, Tanaka, Furuta,
Honda, & Kurosu, 2000), whereas Asahi et al. (1992) researched an automatic
GUI evaluation system. Additionally, Sugizaki (1999) reported on trials of a uni-
fied measuring system for evaluation results; S. Hori, Shimizu, and Nakatani
(1997) described an evaluation method by graph theory for screen layout; Tahira
(2002) reported a determination method by logical precedence with graph theory
for evaluation results, and Murakami, Sano, and Inoue (2002) reported on the
characteristics of evaluation using an eye camera with regard to cost-effective-
ness and persuasiveness.

In recent years, Web site log analysis methods have been proposed as usability
evaluation methods. Fukaya (1999) performed evaluation by combining logging
analysis and state transition diagrams. Several reports have also been published on
Web site evaluation methods that combine analysis of access patterns in the log or
similarity of page content with analysis of quantitative usability evaluation from
eye-tracking data (H. Kato, Nakayama, & Yamane, 2000; Nakayama, Kato, &
Yamane, 2000; Shibuta et al., 2001).

New perspectives on usability evaluation have also been reported. Kuramochi and Ota (2002) organized usability problems from the point of view of learning effect and unusability for skilled users and proposed the Initial-Lasting Matrix (IL-matrix) approach to the importance of problems and the learning effect. Furuta, Sato, and Ogata (2000) reported on a usability evaluation of consumers at a store based on the knowledge that testers might evaluate products positively through performing various tasks. Ogino et al. (2000) reported that usability evaluation was more important than questionnaire use in determining the acceptance and operability of software. Finally, Nakagawa and Mitome (2000) proposed a method to deduce user benefits of products.

2.4 Research Trends in Design and Devices

The first concrete steps toward the introduction of the concept of usability into product design and device development began in the early 1990s, and their results first became public around 1993. Practical application began in the copy machine industry, and the announcement of Windows 3.1 triggered the switch from Character User Interface (CUI) to GraphicalUser Interface (GUI), which was deployed into industrial line-of-business applications. After 1995, as the PC as we know it and display-mounted devices became increasingly popular, problems of operability were noticed in various industries. Until about 2000, practical research into these problems were reported by key staff in each industry (generally project leaders with budgetary responsibilities who were involved in introducing usability into product development). Specific cases regarding Web applications were published from 2001.

Because operability had been the focus of considerations in the copy machine industry since early on, Morozumi, Nagasaki, and Shibata (1994) organized and categorized recognizable patterns based on "backward operation" (i.e., operations not classified as ordinary forward operations), and Etani, Kudoh, Satoh, Itoh, and Fujino (1994) proposed a system of "Communication Engineering" to consider natural information interchanges by reviewing them from the point of view of human-centered design. This report detailed one of the first applications of the concept of human-centered design, proposed in ISO 13407.

Urokohara, Fukuda, Nanami, and Okutani (1996) presented a case that realized considerable cost reductions in design after thorough review and evaluation of operability by utilizing operational flow diagrams in the semiconductor production control systems' user interface, and Hata, Morimoto, Kurokawa, Inoue, and Kushiro (1996) verified the effectiveness of the "Parallel Design Method" proposed by Nielsen (1993) in the operational control panels of building lighting and air conditioning systems.

In the field of home electronic appliances, and with the need to adapt to the high-tech functions and multifunctionality of home information appliances as a result of the spread of the Internet, Kimura, Moriuchi, Kato, and Inokuchi (1998) reported on a video remote control interface that is adaptable to user characteristics, and Hiruma et al. (1999) reported the trial production and evaluation of remote controls for digital broadcasting receivers. Masui, Okazaki, Asano, and Tonomura (1998) pointed out that it was also important to consider the facsimile menu method with regard to display limitations posed by cost factors. Recently, regarding the legibility of Japanese characters displayed in computer screens, Yoshitake et al. (2001) proposed characteristics for more legible characters, taking into account character matrix elements as well as character size.

Looking at device trends, with the increasing opportunities to handle information in the car, such as in car navigation systems, Sakaguchi, Higuchi, Nakano, and Yamamoto (1997) studied improvements in the visibility of car-mounted displays based on the adaptability of the human eye, and Yoshitake et al. (2002) indicated the improvements in usability and specific benefits for users of considerably widening the viewing angle of LCD for notebook PCs.

Saki, Tanigichi, Tamura, and Shibuya (1997) studied menu layout and its usability in action interfaces where users do not need specialized video game input devices. Nishiki, Hatashi, Muroaka, Kawakami, and Fujita (1998) reported the introduction of a switch to negate the flaws in touch panel. Tokunaga et al. proposed not only switch structure but also cost down method in the development and evaluation of effectiveness of double-stroke input devices enabling space-saving design by assigning multiple functions to one operating portion. Hatakeyama et al. (1998) reported on the evaluation of system operation at the device consideration stage of an audio signage system for pedestrians, in line with the idea of universal design of barrier-free concepts. Inoue and Tano (1999) addressed one of the weakest points of the GUI system, the issue of browsability, by proposing a user interface for independent LCD tablets and a set of rules for proposed devices to integrate multiple display devices.

Regarding the World Wide Web, Mayuzumi et al. (2001) proposed a community site with usability in mind where companies are also members of that site. According to the proposal, many important points must be considered when building a community site, such as the integration of the three elements of system, design, and management, considerations from the user's point of view, availability of user information, process design, and experience. Chen et al. (2001) proposed a method of creating effective guidelines for visual Web design by stipulating four criteria for suitability for Web use (operational facility, recognizability, visual suitability, and accessibility) and by assessing these in relation to the purpose of the product being designed. Reported guidelines include Global Web Design Guidelines from Japan—International Web (Shinohara & Ueno, 2002) as a Web guideline as well as the GUI guideline by Fukuda et al. (2002) on the current status and improvement of system messages in the GUI environment. Jin, Mayuzumi, Lee, and Choh (2002) strongly argued that accessibility be improved by taking the approach of user engineering to Web design.

2.5 Human Characteristics and Other

Research activities in this field have their inception in research by Omura et al. (1990) that aimed for a method of system design that took into account ease of use and understanding. This article made a forecast that the production system supported by Computer Supported Cooperative Work would be the mainstream of the field in the future and hence considered the problem of mental stress that affects people joining in the cooperative work.

From this time until 1996 many manual-related research reports were submitted. First of all, in Doi (1992), the relation between ease of understanding and knowledge propagation was noted and the creation of a simple manual, called a summary note, for sales engineers was considered. Niikura and Hamada (1993) explained what kind of role a telephone manual should fill and addressed how to make a manual that is easy to use. Regarding ease of understanding of manuals for beginners, Matsumoto, Ueda, Takeda, and Mizunash (1994) indicated that performance would improve if target-oriented structures and definition methods were included in manuals for task-oriented beginners. Yamada and Kojima (1996) aimed to improve the usability of electronic instruction manuals while allowing for expert users.

From about 1995, research into ease of use and understanding of onscreen elements began to appear. Initially, Kurosu and Kashimura (1995) demonstrated that giving the impression of improving the ease of use of an interface (apparent usability) was important. This report noted a strong correlation between apparent usability and aesthetic satisfaction. To analyze the cognitive factors of icons, Morimoto, Kurokawa, Nishimura, and Torii (1994) experimented with subjective evaluation. Nagasaki et al. (1996) showed how poorly thought out terminology could be a hindrance to the successful operation of office appliances. Kurokawa, Kuze, and Morimoto (1996) described the characteristics of pointing operations on touch screen interfaces. To provide users with easy-to-understand guidance, Aoyama, Ando, and Kurosu (1998) investigated what the factors of guidance are and how those factors affect users.

On the general topic of usability, Kurosu (1996) examined its basic value to human interface specialists. Examples of practical activity in this area include the research of Ando, Aoyama, and Kurosu (1998) into the effects of map deformation on the ease of understanding and that of Wakamatsu and Kurosu (1999) into developing an elevator control panel that is easy to use from a wheelchair; while utilizing the inevitable time lags in a system, Sadakata (1999) carried out trials to reduce the memory burden on users of systems by showing the past operation logs as a guidance.

From about this time, the number of reports concerning user models began to increase. Previously, Morozumi, Nagasaki, and Shibata (1993) had considered operational problems caused by the overlap between the system model and the user's model and had concluded that it was essential to include a design based on the user's model in the first stage of system design. In addition, Nishikawa et al. (1999) proposed a framework they termed "functional adaptation" for progressively improving the interface based on a user cognitive model and user requirements. More recently, Wada, Iwata, and Tano (2000) offered the basic architecture of a model, adaptable to the characteristics of individual users, for presenting information and estimating the user's cognitive load.

From 1999, there was a drastic increase in the number of usability research reports containing the keywords *beginners, the elderly,* and *the handicapped.* Regarding the elderly, Aoyama, Ando, and Kurosu (1999) focused on a previously neglected area of behavioral research in usability testing and deduced a mental model for tools and appliances for the elderly, and investigated the influence of the cognitive process on errors and usability. In researching household electronic products used by the elderly, T. Hori et al. (1999) looked at the differences in operability by age group, designed a remote control with the elderly in mind, and concluded that particular attention must be paid to toggle switches and screen transitions. Further, Takemoto et al. (1999) proposed a new design for remote controls taking into account the characteristics of the elderly, and demonstrated the conditions that a remote control must meet to be user-friendly for the elderly. Taking the theme of a mental model of the elderly, Watanabe and Kurosu (1999) showed how, compared to people in their fifties and sixties, the ability of people in their seventies to operate household appliances deteriorated. Ando, Yoneda, and Ohkubo (2002) analyzed the use of Web sites by the elderly and the visually disabled, and reiterated the importance of designing Web sites so that they are easy to use and navigate. Nambu et al. (2002) carried out research that demonstrated characteristics in how the elderly use IT appliances, while Akatsu et al. (2002) reported the results of testing how the elderly used a video game machine and a high-tech rice cooker. In the latter report, it was shown that middle-aged and elderly people have difficulty in deciding quickly which information is important in using an appliance. Sawashima, Harada, Akatsu, Nambu, and Ishimoto (2002) reported on using a questionnaire method to examine the relation between user attributes and usability testing results, drawing the conclusion that, although the number of features of appliances increased, they can use them easily if they know about the appliances themselves. This is due to the forced application of existing mental models.

Regarding usability for beginners, Kimura, Kato, and Inoue (1999) issued research results proposing methods of customizing consumer electronic appliances to suit individual user characteristics. I. Okamoto and Moriguch (2002) found that, in software for creating homepages, the provision of wizards offered more effective assistance for beginners than the method of clicking commands in menus. Approaching the issue from the idea that simply becoming familiar with a product will make that product easier to use, Kuramochi and Ota (2002) found that the level of expertise of a user did not necessarily increase with the length of time spent using the product, and whether that will grow into a lasting barrier or not will depend on the nature of problems regardless of the user's skill or the length of time used.

In other areas of research, T. Kato, Shirai, and Fukuda (1997) evaluated generic application software featuring the latest GUIs using a questioning protocol method, and T. Kato et al. (1998) considered the sense of control of users of automatic processing functions, concluding that many users are not able to use such functions. In addition, Furukawa and Inagaki (1999) reported on the lack of assistance provided to users of automated systems, whereas Shioyama, Morimoto, and Kurokawa (1998) sought the development of an integrated interface that gave vibratory as well as aural and visual stimuli. Tsukigase and Kurosu (2000) researched online shopping, where there exists a significant discrepancy between ease of use and sense of satisfaction, and Ozaku, Kono, and Kidode (2001) reported on user-centered design from the rare viewpoint of the design and change of a system's internal structure. Finally in this section, K. Takahashi, Nakatani, and Nishida (2002) experimented in information presentation methods with regard to the user's sense of security.

3. USABILITY RESEARCH TRENDS BY AREA

This chapter describes recent research trends based on areas of primary research interest.

3.1 Instruction Manuals and Reference Manuals

The Japan Technical Communicators Association (established in 1992) regularly issues a variety of study results. Of unique interest is a report into video manuals.

3.2 The Web

There has been a lot of research activity in this field in recent years. State transition diagram and log analysis (Fukaya, 1999) and questionnaires (Nakagawa et al., 2001) are now under investigation as usability evaluation methods. Regarding the development processes of Web sites, Mitarai (2001) described cases of improving such processes by direct interaction with real users; Hasegawa and Kobayashi (2001) analyzed the cost and effect of various usability testing methods, and I. Okamoto et al. (2001) reported activities for the long-term improvement of specific Web authoring tools. Regarding specific Web sites, Itoh et al. (2001) analyzed local government sites, and Yoshida, Hatta, Urawa, and Miyanami (1999) analyzed an organization site. Itoh et al. also analyzed company participation-type community sites on the Web; Tsukigase and Kurosu (2000) analyzed online shopping trends, and Ozaku et al. (2001) analyzed a help system for online information retrieval.

3.3 Software

Okada and Asahi (2001) developed an automated tool to evaluate the degree of compliance with GUI standards. Usability evaluation was performed for software such as CAD, CAE (Miyoshi & Yagawa, 2000), word processing (Ogino et al., 2000), image processing (Yoshitake & Hosokawa, 2001), and household accounting (Murakami et al., 2002) using methods such as questionnaires, think aloud, walkthrough, interviewing, and eye tracking.

3.4 Home Electronic Appliances

As home electronic appliances become more and more versatile and multifunctional, the significance of their usability increases. Usability has been verified for devices such as videocassette recorders (Kuramochi & Ota, 2002; Wakizaka, Morimoto, & Kurokawa, 2001), audio-video devices such as image presentation cameras (Waida, 2002), and remote controls for digital broadcast receivers (Hiruma et al., 1999). Kimura et al. (1999) proposed an interface for the next generation of home electronic appliances whereby multiple appliances can be operated by a single controller. Aoyama et al. (1999) and Watanabe and Kurosu (1999) also studied usability from the aspect of mental models of users.

3.5 Auto Vehicle and In-Vehicle Products

Nakagawa and Mitome (2000) developed a method to deduce the ease of use desirable for products as well as the user benefits acquired when that level of ease of use is met. They then applied this method to analyze the usability of automobiles. Urokohara et al. (1999) developed a method to compare the difference in operation times between designers and novice users and applied it to the case of car navigation systems. Tahira (2002) studied the usability problems of the equalizer-adjusting feature of in-vehicle audio devices by applying a structure modeling method employing graph theory.

3.6 Health Care and Public Welfare

Activities are reported in addressing usability problems faced by senior citizens in our aging society. Hashimoto and Watanabe (1999) described educating university students in urban policy from the point of view of aged people by having the students simulate the mental and physical conditions of aged people via a Senior Simulator. Nambu et al. (2002), Akatsu, Harada, Nambu, Sawashima, and Ishimoto (2002), and Sawashima et al. (2002) studied the characteristics of aged people by performing usability testing on IT devices such as telephones and video gaming devices. In other cases, Igeta (2001) studied the usability of X-ray diagnostic systems, and Harada and Shikano (2002) studied the usability of infusion pumps.

3.7 Portable Phones, PDAs, and Mobile Devices

Mine et al. (2001) investigated the user requirements of PDAs by user observation and questionnaire data. Hosono et al. (2002) proposed a design obtained by sensory evaluation based on PDA investigation of both Japanese and foreign subjects.

3.8 Other Areas

There are other areas of interest, such as that of "Process Control," where the control rooms of industrial plants and factory automation systems are studied, and that of "Public Service," where public facilities such as elevators are looked at.

4. ACTIVITIES IN RELATED BUSINESSES

Table 1 gives examples of just a few related activities in Japanese businesses. According to this information, activity actually began in the 1980s, originally in company departments such as design. Evaluation activities such as usability testing be-

Table 1: Examples of Usability Activities in Japanese Businesses

Company	Usability Department	Number of Personnel	Activity	Detail	Summary
IBM Japan	Human Factors Department	5	Usability evaluation of computer-related products (VDT terminals, printers, peripheral devices, personal computers, portable computers, software) and investigation and experimentation of design guideline proposals.	With a large percentage of product development taking place on a global scale, cooperation in Europe and the United States was often required. For this reason, thorough documentation process was already established by the 1980s. Our system is part of the product development process and involves the creation of a Usability Plan on which the subsequent creation of a Usability Test Plan is based. In the 1980s, usability testing was reinforced under the term UOST (User-Oriented System Test), which was concretized as task-based evaluation. From the mid-1990s, UCD was also defined in the product development process, and usability testing constitutes one part of those activities.	Established as an independent section in 1985 and installed a usability laboratory. For subsequent details, refer to the Detail column.
Fuji Xerox	Usability Design Group, Human Interface Design Development Section	10+	Usability evaluation of office appliances and instruction manuals. Investigation of solutions to usability problems, and creation of requirements specifications. Creation of design guidelines. Evaluation and design for universal design.	Positioned, in terms of evaluation, in the quality evaluation process of all Fuji Xerox development activities and charged with evaluating the usability of all development products. The department also has usability design functions.	Evaluation of product usability started in 1983; Usability Evaluation Group established in design organization in 1984. Testing laboratories established in 1991 and 1997.

(continued)

Table 1: Examples of Usability Activities in Japanese Businesses (Continued)

Company	Usability Department	Number of Personnel	Activity	Detail	Summary
Fujitsu Info Software Technologies	Quality Assurance Department	4	Usability testing of Fujitsu products and services, such as software and Web sites. Activities include usability evaluation of all kinds of product, research into business expansion, and public relations activities.	Usability laboratory was established in 2000 as the third laboratory in Fujitsu Group. Based on the know-how of previous laboratories, its design features a round table for discussions with usability testers. As Fujitsu IST is a company that focuses exclusively on software development, the goal of improving usability was accepted with no objection at board level. Each product must pass each of three rounds of usability testing before it can proceed to the next level of inspection.	Usability laboratory established in 2000; in 2002, usability testing was incorporated into development procedures for all company products.
Brother Industries	Distributed throughout the company	6	Canvassing user opinions and performing usability testing for trial products and simulations for Brother products.	At Brother Industries, usability testing is performed by the Design department. With introduction of Company System, usability evaluation method is transferred to the department in charge in the Companies. Companies cooperate on usability evaluation.	Usability laboratory established in 1996, evaluation started. Evaluation overseas started in 1997. Joint evaluation between Head Office and Companies started in 2002.

Company	Division	Staff	Activities	Approach	History
Ricoh	Appliance Promotion Office, CSM Division	22 (Usability Assessment Group: 10, Process Development Group: 7, Planning & Research Group: 2, Other: 3)	Co-operation with Planning, Design and Quality Control Department of Development Division and Design Department to carry out investigation and evaluation of usability. Usability evaluation of products (manufactured by Ricoh and by other companies). Creation of tools. Education on usability issues. Development of human resources. Process evaluation. Analysis of user requirements.	Since ten years ago, usability evaluation has been performed at the development stage by cooperation between architects and designers. Employees registered with our in-company monitor are used as the subjects of evaluation.	MMC Promotion Office inaugurated in 1990; testing laboratory established in 1993; Appliance Promotion Office established in CSM Head Office in 1998.
U'eyes novas Inc.	Usability R&D	13 (Usability Evaluation: 5, UCD Development: 6, User Information Research: 2)	Focused on usability business. Covering fields include requirement definition, basic specifications design, operational flow design, visual interface design for production phase. With professionals of cognitive psychology and ergonomics, evaluation plans for problem solution are scheduled and performed. Other businesses are holding usability seminars, consulting of process improvements for organizations, and development & sales of evaluation analysis tools.	For all the product development processes, know-hows exist to improve operability. Many original know-hows include professional evaluation by the staffs with rich experience and persuasive quantitative NEM evaluation method. Features large number of testers of 1000 people, contextual user information collection, statistics analysis, wide range of support product fields for user-centered development such as everyday products, public devices, and car cockpits.	User interface design business started in 1991; professional evaluation started in 1993; user testing started in 1997; user testing laboratory installed in 1999; addition of one more laboratory planned in 2003.

came prominent in the 1990s, gradually finding their place at the heart of the entire development process. Although the number of personnel involved widely ranges from a handful to over 20, this is related to the scope of the department's work. Naturally this work consists largely of usability evaluation, but not exclusively. Other activities include creating design guidelines and requirement specifications, in-company PR activities, and human resources development.

5. CONCLUSIONS

This article has presented a general outline of usability research activities and practical activities in Japan, with a particular focus on the last ten years. Although there have been research activities carried out by universities and other research organizations, greater effort and advancements are to be seen in the work of related businesses. In Japan, the Human Interface Society contains a Special Interest Group on Usability (http://sig.his.gr.jp/usability/) that holds regular meetings for interested parties in the Tokyo region to publish research and exchange information. With the acceleration in the publication of related material over the last few years, it is safe to conclude that usability activity in Japan is in a vigorously healthy state. Although an "English barrier" has prevented activities in Japan from becoming widely known abroad, if this article can contribute to the bridging of this information gap, the authors will be more than happy.

REFERENCES

Akatsu, A., Harada, E. T., Nambu, M., Sawashima, H., & Ishimoto, A. (2002). Relational analysis of usability testing on information devices (2): How elderly people use a video-game machine and a high-tech rice cooker? *Proceedings of the Human Interface Symposium 2002, 2113,* 199–200.

Amemiya, T. (1990). Analyzing the structure of understandability and pleasantness of interface experiences: A view from mediated interactionism. *Proceedings of the Human Interface Symposium 1990,* 291–298.

Ando, M., Yoneda, K., & Ohkubo, A. (2002). Analysis of Web use characteristics of the elderly and the visually disabled. *Proceedings of the Human Interface Symposium 2002, 3213.*

Aoyama, K., Ando, K., & Kurosu, M. (1998). The usability factors of real time guidance. *Proceedings of the 14th Symposium on Human Interface,* 23–28.

Aoyama, K., Ando, K., & Kurosu, M. (1999). Developmental process of the mental model in the usability testing situation. *Proceedings of the Human Interface Symposium 1999,* 837–842.

Arai, N. (2001). The practice in improvement of the usability of "isize.com". *Correspondences on Human Interface, 3*(4), 9–10.

Arai, N., Itoh, J., Urokohara, H., & Takamoto, Y. (2002). Practical usability activities in corporations. *Journal of Human Interface Society, 4*(1), 25–34.

Araki, S., Sugizaki, M., Itoh, Y., & Kurosu, M. (2001). On the analytical method of the context of use for the purpose of generating adequate product concepts—3. Revised method of the sequence model. *HIS 2001,* 545–548.

Bayer, H., & Holtzblatt, K. (1998). *Contextual design.* London: Morgan Kaufmann.

Bevan, N., Earthy, J., & Kurosu, M. (2001) The benefits of using ISO13407: Human centred design processes for interactive systems. *INTERACT2001*, p. 843.

Carroll, J. M. (Ed.). (1995). *Scenario-based design*. New York: Wiley.

Chen, L., Jin, H., Lee, N., Yamashita, S., Ito, K., Mayuzumi, Y., & Choh, I. (2001) The proposal of Web design utilizing visual usability efficiently. *Proceedings of the Human Interface Symposium 2001*, 333–334.

Cooper, A. (1999). *The inmates are running the asylum. Sams Programming* (H. Yamagata, Trans.). Tokyo: Shoeisha.

Doi, M. (1992). Summary note creation based on novice users experiments. *Progress in Human Interface, 1*, 27–34.

Etani, N., Kudoh, H., Satoh, K., Itoh, M., & Fujino, T. (1994). Applying human-behavior analysis and design of man-machine interaction. *Progress in Human Interface, 3*(1), 23–29.

Fukaya, M. (1999). Evaluating Web usability with state transition diagram and log analysis. *Proceedings of the Human Interface Symposium 1999*, 525–530.

Furukawa, H., & Inagaki, T. (1999). A dynamic abstraction hierarchy as a basic framework for state representation of automated systems. *Proceedings of the Human Interface Symposium 1999*, 27–32.

Furuta, K., & Sato, D. (2001). A check-boxed observation sheet method—Pre-inspected user testing. *Proceedings of the Human Interface Symposium 2001*, 379–382.

Furuta, K., Sato, D., & Ogata, S. (2000). Do-gugan: Usability judging sight—Consumers' initial judgement of usability. *Proceedings of the Human Interface Symposium 2000*, 321–322.

Harada, E. T., & Shikano, Y. (2002). Usability testing under pressure: A testing method for an expert-oriented device. *Proceedings of the Human Interface Symposium 2002*, 143–146.

Hasegawa, A., & Kobayashi, N. (2001). Usability testing in a Web site development. *Correspondences on Human Interface, 3*(4), 1–4.

Hata, M., Morimoto, K., Kurokawa, T., Inoue, M., & Kushiro, N. (1996). Parallel design of a control panel and its usability testing. *Proceedings of the 12th Symposium on Human Interface*, 553–558.

Hatakeyama, T., Ito, K., Shiratori, T., Shiroguchi, M., Kurachi, K., & Kasuga, M. (1998). Audio signage system for pedestrians. *Proceedings of the 14th Symposium on Human Interface*, 577–582.

Hiruma, N., Morita, T., Komine, K., Ishiyama, K., Ito, T., Isono, H., et al. (1999). Design and evaluation of remote controllers for digital broadcasting receivers. *Proceedings of the Human Interface Symposium 1999*, 489–492.

Hori, S., Shimizu, Y., & Nakatani, T. (1997). Study of evaluating method of screen construction with oriented graph. *Proceedings of the 13th Symposium on Human Interface*, 523–526.

Hori, T., Takemoto, K., Morimoto, K., Kurokawa, T., Kushiro, N., & Inoue, M. (1999). Design of remote controllers of household electric appliances for the elderly and their usability testing . *Proceedings of the Human Interface Symposium 1999*, 261–266.

Hosono, N., Inoue, H., Tomita, Y., & Yamamoto, Y. (2002). Human-centered design by utilizing sensory analysis—To compare the requirements from Japanese subjects with others, and to verify by the experts. *The Transactions of Human Interface Society, 4*(3), 175–180.

Igeta, Y. (2001). Usability design and evaluation of x-ray diagnostic system. *Correspondences on Human Interface, 3*(4), 21–24.

Inoue, S., & Tano, S. (1999). Design of user interface for littery augmented papers (LAP). *Proceedings of the Human Interface Symposium '99*, 17–22.

Itoh, Y., Ikeya, Y., & Nakano, M. (2001). Current status and problems of public Web sites based on the usability evaluation of twenty-eight Japanese cyukaku-city (Large city with much of independence) Web sites. *Correspondences on Human Interface, 3*(4), 15–20.

Ito, Y., Yamada, F., & Kurosu, M. (2000). Quantitative and qualitative analysis of the effects of understanding and modeling of the fieldwork data over the generation of new system concepts. *Correspondences on Human Interface, 2*(4), 7–12.

Jin, H., Mayuzumi, H., Lee, N., & Choh, I. (2002). The study of Web designing with approach of user engineering—A proposal on Web design in consideration of visually handicapped person's accessibility. *Proceedings of the Human Interface Symposium 2002*, 483–486.

Kajii, K., & Itoh, M. (2001). How is product centered testing different from "usability" evaluation in HCD? *Proceedings of the Human Interface Symposium 2001*, 409–412.

Kanayama, T., Ogasawara, H., & Kimijima, H. (2002, June). 7th European Conference on Software Quality, Helsinki, Finland.

Kato, H., Nakayama, T., & Yamane, Y. (2000). *Navigation analysis tool based on the correlation between contents distribution and access patterns.* WebKDD2000.

Kato, T., Ikemura, S., Tatsuno, M., Toyozawa, S., Niwa, M., Murai, A., et al. (1998). A trade-off between automating cumbersome procedures and assuring the internal locus of control. *Proceedings of the 14th Symposium on Human Interface*, 49–54.

Kato, T., Shirai, K., & Fukuda, M. (1997). When should syntactic consistency be violated? *Proceedings of the 13th Symposium on Human Interface*, 505–510.

Kawasaki, T., Kashimura, K., & Harada, E. (2000). The effect of multiple participants in usability testing—Co-participation method vs. Thinking aloud method. *Proceedings of the Human Interface Symposium 2000*, 347–350.

Kimura, A., Kato, H., & Inoue, S. (1999). Next generation user interface for consumer devices—Fundamental study of customizable interfaces. *Proceedings of the Human Interface Symposium 1999*, 493–498.

Kimura, A., Moriuchi, M., Kato, H., & Inokuchi, S. (1998) A study on customized remote control interface based on user characteristics—Needed user characteristics to design video remote-controllers. *Proceedings of the 14th Symposium on Human Interface*, 43–48.

Kitajima, M., Blackmon, M. H., Polson, P. G., & Lewis, C. (2002). AutoCWW: Automated cognitive walkthrough for the Web. *Proceedings of the Human Interface Symposium 2002*, 271–274.

Kuramochi, A., & Ota, S. (2002). Learning effect and obstructions—Is it true "It is all right if it gets accustomed?" *Proceedings of the Human Interface Symposium 2002*, 281–288.

Kurokawa, T., Kuze, M., & Morimoto, K. (1996). Characteristics of human pointing operation in touch screen interface and applicability of Fitts' law to them. *Progress in Human Interface, 5*, 5–12.

Kurosu, M. (1996). The structure of the Usability Concept. *Human Interface News & Report, 11*, 351–356.

Kurosu, M. (1997a). Dilemma of usability engineering. *HCI International '97, 21B*, 555–558.

Kurosu, M. (1997b). Experimental validation of DTM (dual task model). *ASEAN Ergonomics '97*, 419–423.

Kurosu, M. (1998). Eye-hand coordination in the dual task situation. *Global Ergonomics Proceedings*, 657–660.

Kurosu, M. (2001). Globalization and localization—Structure of the diversity. *HCI International 2001, 2*, 609–611.

Kurosu, M. (Ed.). (2003). *Usability testing.* Japan: Tokyo: Kyoritsu-shuppan.

Kurosu, M., Ito, M., Horibe, Y., Hirasawa, N., & Fujiwara, Y. (2000, August). *Diagnosis of human-centeredness of the design process by the SDOS.* Usability Professional's Association 9th Annual Conference, Asheville, NC.

Kurosu, M., Ito, M., & Tokitsu, T. (1999). *User engineering.* Kyoritsu-shuppan.

Kurosu, M., Sugizaki, M., & Matsuura, S. (1998). Structured heuristic evaluation (sHEM). *Usability Professionals' Association 7th annual Conference Proceedings*, 3–5.

Kurosu, M., Urokohara, H., Sato, D., Nishimura, T., & Yamada, F. (2002). A new data collection method for usability testing: NEM (Novice Expert ratio Method). *Usability Professionals' Association 11th annual Conference Proceedings, No. 51.*

Landauer, T. K. (1995). *The trouble with computers* (H. Yamagata, Trans.). Cambridge, MA: MIT Press.

Masui, N., Okazaki, T., Asano, Y., & Tonomura, Y. (1998). A study of facsimile menu in small display area. *Proceedings of the 14th Symposium on Human Interface, 485–492.*

Matsumoto, F., Ueda, Y., Takeda, K., & Mizunashi, S. (1994). Understanding-oriented manual vs operation-oriented manual. *Proceedings of the 10th Symposium on Human Interface,* 317–322.

Matsuura, S., Kurosu, M., & Sugisaki, M. (1998). Experimental approach to cognitive walkthrough method I. *Proceedings of the 14th Symposium on Human Interface, 29–32.*

Mayuzumi, Y., Ito, K., Yamashita, S., Riwen, C., Jeong, J., & Choh, I. (2001). Study of communication design for ecological site in Japanese company—Trial of problems for composing and soluble technique by visual representation in ecological site. *HIS 2001, 327–330.*

Mine, J., Ikeda, T., Inagaki, K., & Okuizumi, N. (2001). Survey of user requirements and its approach. *Proceedings of the Human Interface Symposium 2001, 413–416.*

Mitarai, D. (2001). Case study of Web site improvement by direct interaction with site users. *Correspondences on Human Interface, 3*(4), 5–8.

Mitsubishi Electric Corporation Industrial Design Center (Ed.). (2001). *These design will give us the usability.* Tokyo: Kougyou-chousa-kai.

Miyoshi, A., & Yagawa, G. (2000). The usability of CAD and CAE. *Correspondences on Human Interface, 2*(4), 25–26.

Morimoto, K., Kurokawa, T., Nishimura, T., & Torii, T. (1994). Analysis of cognitive factors of icons based on subjective evaluation. *Progress in Human Interface, 3,* 65–72.

Morozumi, K., Nagasaki, M., & Shibata, K. (1993). Analysis of the interaction on backward operations of batch type equipment. *Proceedings of the 9th Symposium on Human Interface,* 479–484.

Morozumi, K., Nagasaki, M., & Shibata, K. (1994). Analysis of the interaction on backward operations of batch type equipment. *Progress in Human Interface, 3*(2), 73–80.

Murakami, Y., Sano, M., & Inoue, M. (2002). Researching the methodology of usability testing using eye-camera—Its efficiency and cost effectiveness. *Proceedings of the Human Interface Symposium 2002, 293–296.*

Nagamatsu, T., Yamamoto, M., Iwata, D., & Yoshikawa, H. (2001). An experimental study on user-centered design and evaluation for a cooperative machine maintenance training system based on virtual environment. *Journal of Human Interface Society, 3*(4), 307–318.

Nakagawa, K., & Mitome, S. (2000). Development of methodology to determine product/service usability and user benefits. *Correspondences on Human interface 2000/11/22, 2*(4), 1–6.

Nakagawa, K., Suda, T., Zempo, H., & Matsumoto, K. (2001). The development of questionnaire for evaluating Web usability. *Proceedings of the Human Interface Symposium 2001,* 421–428.

Nakajima, K., Ichikawa, T., & Kato, T. (2002). Designing a supporting software tool for cognitive walkthrough. *Proceedings of the Human Interface Symposium 2002, 625–628.*

Nakakoji, K., Kamiya, T., Takashima, A., & Yamamoto, Y. (2001). A representation and interactive viewers for human-computer interaction. *Proceedings of the Human Interface Symposium 2001, 155–158.*

Nakayama, T., Kato, H., & Yamane, Y. (2000, May). *Discovering the gap between Web site designers' expectations and users' behavior.* 9th International WWW conference, Amsterdam, Holland.

Nambu, M., Harada, E. T., Akatsu, H., Sawashima, H., & Ishimoto, A. (2002). Relational analysis of usability testing on information devices (1): How elderly people use automated teller machine and L-mode telephone. *Proceedings of the Human Interface Symposium 2002*, 195–198.

Nielsen, J. (1993). *Usability engineering* (M. Shinohara, Trans.). Toppan, Tokyo: Denki University Press.

Niikura, Y., & Hamada, Y. (1993). Re-examining the role of the telephone operation manual —An experiment incorporating user and task perspective. *Proceedings of the 9th Symposium on Human Interface*, 471–474.

Nishikawa, M., Kuramochi, Y., Takahashi, M., Matsuyama, S., Fujisawa, M., & Kitamura, M. (1999). Functional adaptation of MMI based on cognitive model and user requirements— Application to experimental accelerator facility. *HIS 1999*, 825–830.

Nishiki, T., Hatashi, T., Muraoka, Y., Kawakami, M., & Fujita, T. (1998). Usability improvement in television broadcasting studio by implementing CC switch. *Proceedings of the Fourteenth Symposium on Human Interface*, 55–60.

Norman, D. A. (1988). *The psychology of everyday things* (H. Nojima, Trans.). New York: Basic Books.

Ogino, M., Ichikawa, M., & Shimamura, R. (2000). Usability study on the new product development process from the viewpoint of human-centered design and future work. *Correspondences on Human interface 2000/11/22*, 2(4), 21–24.

Ohtsuka, I. (1992). Organizing user-interface design principles from "user centered" viewpoint. *Proceedings of the Human Interface Symposium 1992*, 571–576.

Okada, H., & Asahi, T. (2001). Development of an automatic GUI design checking tool for design standards. *The Transactions of Human Interface Society*, 3(1), 1–8.

Okamoto, I., & Moriguch, K. (2002). A study on effective user assistance for home use software. *Proceedings of the Human Interface Symposium 2002*, No. 2232.

Okamoto, I., Tanaka, S., & Itoh, H. (2001). Continuous usability improvement on designing Web authoring tool. *Proceedings of the Human Interface Symposium 2001*, 417–420.

Okamoto, T., & Kurosu, M. (2000). A questionnaire to evaluate the subjective usability. *Proceedings of the Human Interface Symposium 2000*, 25–28.

Omata, M., Tanoue, K., Go, K., & Imamiya, A. (2000). Developing a remote usability evaluation system on information networks—Developing a virtual test room. *Correspondences on Human Interface 2002/2/21*, 2(1), 1–6.

Omura, H., Kuzuoka, H., Warisawa, S., Mitsuishi, M., Hirose, M., & Ishii, T. (1990). A study on evaluation of CSCW in manufacturing systems, considered from a human engineering approach. *Proceedings of the 6th Symposium on Human Interface*, 105–108.

Oonishi, J., & Go, K. (2002). *Requirement engineering*. Tokyo: Kyoritsu-shuppan.

Ozaku, H., Kono, Y., & Kidode, M. (2001). A study of the usability for the assist system to make tourist routes. *Proceedings of the Human Interface Symposium 2001*, 429–432.

Rosson, M. B., & Carroll, J. M. (2002). *Usability engineering—Scenario-based development of human-computer Interaction*. London: Morgan Kaufmann.

Sadakata, T. (1999). Interface for a system that includes inevitable time lags—Printing ink-key control interface showing user behavior. *Proceedings of the Human Interface Symposium 1999*, 69–72.

Sakaguchi, Y., Higuchi, K., Nakano, T., & Yamamoto, S. (1997). Visibility improvement of on-board display based on the adaptation function of eyes. *Proceedings of the 13th Symposium on Human Interface*, 527–532.

Saki, T., Taniguchi, A., Tamura, H., & Shibuya, Y. (1997). On the usability of action interface related to menu allocation. *Human Interface News & Report*, 12 No.(1), 13–18.

Sato, J., Itoh, S., & Kobayashi, T. (2002). Incorporation of repetitive usability testing into software development process. *Correspondences on Human Interface*, 4(5), 65–70.

Sawashima, H., Harada, E. T., Akatsu, H., Nambu, M., & Ishimoto, A. (2002). Relational analysis of usability testing on information devices (3): Relation between interactions with IT devices, and cognitive abilities and/or traits in daily lives. *Proceedings of the Human Interface Symposium 2002, 201–202.*

Shibata, Y. (1992). User-participation to design human-centered systems. *Proceedings of the Human Interface Symposium 1992, 585–592.*

Shibuta, K., Ohyama, T., Sakai, K., Yuzawa, H., Mihira, T., Hirose, Y. et al. (2001). An evaluation method of Web site design. *Japanese Society for the Science of Design.*

Shinohara, T., & Ueno, M. (2002). Global Web design guidelines from Japan—International Web. *UPA2002 annual conference proceedings, 25.*

Shioyama, A., Morimoto, K., & Kurokawa, T. (1998). Subjective correspondence among duration of vibratory stimulus to a hand and visual and auditory variables. *Proceedings of the 14th Symposium on Human Interface, 499–504.*

Sugizaki, M., Araki, S., Itoh, Y., & Kurosu, M. (2001). On the analytical method of the context of use for the purpose of generating adequate product concepts—2. Ambiguity of the flow model. *HIS 2001, 541–544.*

Suzuki, S., Kaneko, A., & Ohmura, K. (1995). Usability evaluation system QUIS. *Proceedings of the 11th Symposium on Human Interface, 747–752.*

Tahira, H. (2002). Usability problem solving with graph theory—A case study in introducing structure modeling. *Proceedings of the Human Interface Symposium 2002, 277–280.*

Takahashi, K., Nakatani, M., & Nishida, S. (2002). Information presentation from the standpoint of a sense of security. *Proceedings of the Human Interface Symposium 2002, 2243.*

Takahashi, M., Ito, K., Niwa, Y., & Kitamura, M. (1999). Complexity reduction criteria & user requirement specification for interface design. *HIS 1999, 65–67.*

Takemoto, K., Hori, T., Moromoto, K., Kurokawa, T., Kushiro, N., & Inoue, M. (1999). Design of remote controllers based on operation properties of elderly people and evaluation of their ease-to-use. *Proceedings of the Human Interface Symposium 1999, 267–272.*

Takeuchi, K. (1997). Scenario-based usability evaluation and masking method. *Proceedings of the 13th Symposium on Human Interface, 533–538.*

Takeuchi, K., & Miki, H. (1996). Masking method for usability evaluation. *Proceedings of the 12th Symposium on Human Interface, 697–705.*

Tsuchiya, K., Horii, K., Kitamura, Y., & Tamura, H. (1996). User interface in designing interactive computer systems. *Human Interface News & Report, 11(1), 79–86.*

Tsukigase, Y., & Kurosu, M. (2000). A comparative study of on-line shopping and real shopping. *Proceedings of the Human Interface Symposium 2000, 379–382.*

Urokohara, H. (1996). Operation flow charts to straighten up tasks and specifications for programming. *Human Interface N&R 11(1), 75–78.*

Urokohara, H., Fukuda, E., Nanami, H., & Okutani, H. (1996). Examples of user-interface for semiconductors production control systems. *Proceedings of the 12th Symposium on Human Interface, 537–542.*

Urokohara, H., Furuta, K., Tanaka, K., & Kurosu, M. (1999). A usability evaluation method that compare task performance between expert and novice. *Proceedings of the Human Interface Symposium 1999, 537–542.*

Urokohara, H., Tanaka, K., Furuta, K., Honda, M., & Kurosu, M. (2000). NEM: "Novice Expert ratio Method" a usability evaluation method to generate a new performance measure. *ACM SIGCHI2000 Extended Abstracts, 185–186.*

Wada, F., Iwata, M., & Tano, S. (2000). Model to estimate user's cognitive load for adaptive information presentation. *Proceedings of the Human Interface Symposium 2000, 499–502.*

Waida, R. (2002). Human-centered design process for consumer audio-video equipments. *Proceedings of the Human Interface Symposium 2002, 443–444.*

Wakamatsu, M., Sawada, K., & Nomoto, K. (2002). Usability activities in design laboratory of Mitsubishi-electric corporation. *Journal of Human Interface Society, 4*(4), 207–212.

Wakizaka, Y., Morimoto, K., & Kurokawa, T. (2001). A proposal of absolute index for usability of information apparatus. *Proceedings of the Human Interface Symposium 2001,* 425–428.

Watanabe, S., & Kurosu, M. (1999). Study on universal interface of daily-use products. *Proceedings of the Human Interface Symposium 1999,* 273–276.

Yamaoka, T., & Baber, C. (2000). 3 point task analysis and human error estimation. *HIS 2000,* 395–398.

Yamaoka, T., Suzuki, K., & Fujiwara, Y. (Eds.). (2000). *Design and evaluation of structured user interface.* Tokyo: Kyoritsu-shuppan.

Yoshida, J., Hatta, A., Urawa, K., & Miyanami, M. (1999). Web usability in Osaka gas. *Proceedings of the Human Interface Symposium 1999,* 435–438.

Yoshitake, R., & Hosokawa, K. (2001). Effectiveness of applying user-centered design to the development of a consumer image processing software. *Proceedings of the Human Interface Symposium 2001,* 167–170.

INTERNATIONAL JOURNAL OF HUMAN–COMPUTER INTERACTION, *17*(1), 125–126

Schedule of Events
2004

April 20–23, 2004   **Berkeley, California, USA**
CFP '04: ACM's 14th Conference on Computers, Freedom and Privacy 2004
URL: http://www.cfp2004.org/

April 24–29, 2004   **Vienna, Austria**
CHI '04: CHI 2004 Conference on Human Factors in Computing Systems
URL: http://www.sigchi.org/chi2004

May 15–22, 2004   **New York, New York, USA**
WWW'04: The 2004 World Wide Web Conference
URL: http://www.www2004.org

May 25–28, 2004   **Gallipoli, Italy**
AVI'04: International Conference on Advanced Visual Interfaces
URL: http://www.di.uniba.it/~avi2004/

June 7–11, 2004   **Tucson, Arizona, USA**
JCDL '04: ACM/IEEE Joint Conference on Digital Libraries 2004
URL: http://www.jcdl2004.org/

June 15–18, 2004   **Suntec City, Singapore**
GRAPHITE '04: International Conference on Computer Graphics and Interactive Techniques in Australiasia and South East Asia (co-located with VRCAI 2004 Conference)

June 27–July 1, 2004   **Karlstad, Sweden**
CATaC'04: Cultural Attitudes towards Technology and Communication
URL: http://www.it.murdoch.edu.au/catac/

July 18–21, 2004   **Cambridge, Massachusetts, USA**
DIS '04: Designing Interactive Systems 2004
URL: http://www.acm.org/sigchi/dis2004

August 1–4, 2004   **Cambridge, Massachusetts, USA**
DIS '04: Designing Interactive Systems 2004
URL: http://turing.acm.org/sigs/sigchi/dis2004/

August 8–12, 2004   **Los Angeles, California**
SIGGRAPH '04: Special Interest Group on Computer Graphics and Interactive
Techniques
URL: http://www.siggraph.org/s2004/

November 1–30, 2004   **Washington, D.C., USA**
CIKM '04: Conference on Information and Knowledge Management 2004
URL: http://campus.acm.org/calendar/confpage.cfm?ConfID=2004-5772

Contributor Information

The International Journal of Human–Computer Interaction considers scholarly contributions of various types, including:

Original Research: These manuscripts address the cognitive, neural, social, health, and ergonomic aspects of work with computers and emphasize both the human and computer science aspects of the effective design and use of computer interactive systems. The journal presents original research both in the generic aspects of interface design and in the special applications of interface design in a variety of diversified leisure and work activities.

Survey Papers: Review and reappraisal of critical areas of interest to the human–computer interaction (HCI) research and development community. Contributions in this category should provide well-organized summaries of large bodies of existing literature, synthesized into critical insights that extend current theories and practices.

Research Reports: Pilot studies or small sample size experiments that provide preliminary findings that uncover promising new directions in HCI research.

Case Studies: Implementation or development efforts that have insightful lessons learned from design, development, and implementation of design techniques or new interactive technologies.

Industry Trends: These contributions will review "the best" research and practice that major organizations have in HCI, highlighting both successful and ineffective HCI practices.

MANUSCRIPT PREPARATION: Use a word processor to prepare manuscript. Using 8½- × 11-in. nonsmear paper, type all components (a) double-spaced, (b) 1,800 to 2,000 characters per page (70 to 75 characters per line [including spaces] × 25 to 27 lines per page), (c) on one side of the paper, (d) with each component beginning on a new page, and (e) in the following order—title page (p. 1), abstract (p. 2), text (including quotations), references, appendices, footnotes, tables, and figure captions. Consecutively number all pages (including photocopies of figures). Indent all paragraphs.

Title Page and Abstract: On page 1, type (a) article title, (b) author name(s) and affiliation(s), (c) running head (abbreviated title, no more than 45 characters and spaces), (d) acknowledgments, and (e) name and address of the person to whom requests for reprints should be addressed. On page 2, type an abstract (≤ 150 words).

Editorial Style and References: Prepare manuscripts according to the *Publication Manual of the American Psychological Association* (5th ed., 2001; APA Order Department, P.O. Box 2710, Hyattsville, MD 20784). Follow "Guidelines to Reduce Bias in Language" (APA Manual, pp. 61–76).

Double-space references. Compile references alphabetically (see *APA Manual* for multiple-author citations and references). Spell out names of journals. Provide page numbers of chapters in edited books. Text citations must correspond accurately to the references in the reference list.

Tables: Refer to *APA Manual* for format. Double-space. Provide each table with explanatory title; make title intelligible without reference to text. Provide appropriate heading for each column in table. Indicate clearly any units of measurement used in table. If table is reprinted or adapted from another source, include credit line. Consecutively number all tables.

Figures and Figure Captions: Figures should be (a) high-resolution illustrations or (b) glossy, high-contrast black-and-white photographs.

Do not clip, staple, or write on back of figures; instead, write article title, figure number, and TOP (of figure) on label and apply label to back of each figure. Consecutively number figures. Attach photocopies of all figures to manuscript.

Consecutively number captions with Arabic numerals corresponding to the figure numbers; make captions intelligible without reference to text; if figure is reprinted or adapted from another source, include credit line.

COVER LETTER, PERMISSIONS, CREDIT LINES: In cover letter, include contact author's postal and e-mail addresses and phone and fax numbers.

Only original manuscripts will be considered for publication in the *International Journal of Human–Computer Interaction*. The cover letter should include a statement that the findings reported in the manuscript have not been previously published and that the manuscript is not being simultaneously submitted elsewhere.

Authors are responsible for all statements made in their work and for obtaining permission to reprint or adapt a copyrighted table or figure or to quote at length from a copyrighted work. Authors should write to original author(s) and original publisher to request nonexclusive world rights in all languages to use the material in the article and in future editions. Include copies of all permissions and credit lines with the manuscript. (See p. 174 of *APA Manual* for sample credit lines.)

MANUSCRIPT SUBMISSION: Submit one (1) original and two (2) high-quality manuscript copies to the Editor (see address below). IJHCI also requires soft copy submissions in PDF or Microsoft Word format. These can be sent on disc with the paper copies or submitted via e-mail to the Editor. If sent via e-mail, the subject header should read "IJHCI Manuscript Submission."

Dr. Kay M. Stanney
Industrial Engineering/Management Systems
University of Central Florida
4000 Central Florida Blvd.
Orlando, FL 32816–2450
stanney@mail.ucf.edu

MANUSCRIPT REVIEW AND ACCEPTANCE: All manuscripts are peer reviewed.

Authors of accepted manuscripts submit (a) disk containing two files (word-processor and ASCII versions of final version of manuscript), (b) printout of final version of manuscript, (c) camera-ready figures, (d) copies of all permissions obtained to reprint or adapt material from other sources, and (e) copyright-transfer agreement signed by all co-authors. Use a newly formatted disk and clearly label it with journal title, author name(s), article title, file names (and descriptions of content), names of originating machine (e.g., IBM, Mac), and word processor used.

It is the responsibility of the contact author to ascertain that all co-authors approve the accepted manuscript and concur with its publication in the journal.

Content of files must exactly match that of manuscript printout, or there will be a delay in publication. Manuscripts and disk are not returned.

PRODUCTION NOTES: Authors' files are copyedited and typeset into page proofs. Authors read proofs to correct errors and answer editors' queries.

Printed and bound by CPI Group (UK) Ltd, Croydon, CR0 4YY

22/10/2024

01777637-0014